# The IMA Volumes in Mathematics and its Applications

## Volume 112

*Series Editor*
Willard Miller, Jr.

**Springer**
*New York*
*Berlin*
*Heidelberg*
*Barcelona*
*Hong Kong*
*London*
*Milan*
*Paris*
*Singapore*
*Tokyo*

# Institute for Mathematics and its Applications
# IMA

The **Institute for Mathematics and its Applications** was established by a grant from the National Science Foundation to the University of Minnesota in 1982. The IMA seeks to encourage the development and study of fresh mathematical concepts and questions of concern to the other sciences by bringing together mathematicians and scientists from diverse fields in an atmosphere that will stimulate discussion and collaboration.

The IMA Volumes are intended to involve the broader scientific community in this process.

Willard Miller, Jr., Professor and Director

* * * * * * * * * * *

## IMA ANNUAL PROGRAMS

| | |
|---|---|
| 1982–1983 | Statistical and Continuum Approaches to Phase Transition |
| 1983–1984 | Mathematical Models for the Economics of Decentralized Resource Allocation |
| 1984–1985 | Continuum Physics and Partial Differential Equations |
| 1985–1986 | Stochastic Differential Equations and Their Applications |
| 1986–1987 | Scientific Computation |
| 1987–1988 | Applied Combinatorics |
| 1988–1989 | Nonlinear Waves |
| 1989–1990 | Dynamical Systems and Their Applications |
| 1990–1991 | Phase Transitions and Free Boundaries |
| 1991–1992 | Applied Linear Algebra |
| 1992–1993 | Control Theory and its Applications |
| 1993–1994 | Emerging Applications of Probability |
| 1994–1995 | Waves and Scattering |
| 1995–1996 | Mathematical Methods in Material Science |
| 1996–1997 | Mathematics of High Performance Computing |
| 1997–1998 | Emerging Applications of Dynamical Systems |
| 1998–1999 | Mathematics in Biology |
| 1999–2000 | Reactive Flows and Transport Phenomena |
| 2000–2001 | Mathematics in Multi-Media |

Continued at the back

M. Elizabeth Halloran   Seymour Geisser
Editors

# Statistics in Genetics

With 32 Illustrations

Springer

M. Elizabeth Halloran
Department of Biostatistics
Emory University
Rollins School of Public Health
1518 Clifton Road, NE
Atlanta, GA 30322, USA
betz@bear.sph.emory.edu

Seymour Geisser
School of Statistics
University of Minnesota
270b Vincent Hall
206 Church Street SE
Minneapolis, MN 55455-0436, USA

*Series Editor:*
Willard Miller, Jr.
Institute for Mathematics and its Applications
University of Minnesota
Minneapolis, MN 55455, USA

Mathematics Subject Classifications (1991): 62F03, 62F10, 62F15, 62G09, 62G10, 62P10

Library of Congress Cataloging-in-Publication Data
Statistics in genetics / [edited by] M. Elizabeth Halloran, Seymour Geisser.
p. cm. — (IMA volumes in mathematics and its applications ; v. 112)
Includes bibliographical references.
ISBN 978-1-4419-3170-2
1. Genetics—Statistical methods—Congresses. I. Halloran, M. Elizabeth. II. Geisser, Seymour. III. Series.
QH438.4.S73S73 1999
576.5′07′27—dc21 99-18391

Printed on acid-free paper.



Camera-ready copy prepared by the IMA.

Printed in the United States of America.

9 8 7 6 5 4 3 2 1

## FOREWORD

This IMA Volume in Mathematics and its Applications

## STATISTICS IN GENETICS

is one of the series based on the proceedings of a very successful 1997 IMA Summer Program on "Statistics in the Health Sciences."

I would like to thank the scientific organizers: M. Elizabeth Halloran of Emory University (Biostatistics) and Seymour Geisser of University of Minnesota, (Statistics) for their excellent work as organizers of the meeting and for editing the proceedings. I am grateful to Donald A. Berry, Duke University (Statistics); Patricia Grambsch, University of Minnesota (Biostatistics); Joel Greenhouse, Carnegie Mellon University (Statistics); Nicholas Lange, Harvard Medical School (Brain Imaging Center, McLean Hospital); Barry Margolin, University of North Carolina-Chapel Hill (Biostatistics); Sandy Weisberg, University of Minnesota (Statistics); Scott Zeger, Johns Hopkins University (Biostatistics); and Marvin Zelen, Harvard School of Public Health (Biostatistics) for organizing the six weeks summer program.

I also take this opportunity to thank the National Science Foundation (NSF) and the Army Research Office (ARO), whose financial support made the workshop possible.

Willard Miller, Jr., Professor and Director

# PREFACE

This volume contains refereed papers from the participants of the first week of a six-week workshop on Statistics in the Health Sciences held by the Institute of Mathematics and its Applications in Minneapolis, Minnesota, in the summer of 1997. The organizing committee of the workshop decided to start out with statistics applied to micro entities, so the first week was on Genetics, and proceed to macro units, with the sixth week on Epidemiology and Environment.

The purpose of the week on Statistics in Genetics was to bring together researchers who would not necessarily have attended the same meeting to promote productive and novel discussion. The participants ranged from population biologists with strong statistical applications to mathematical statisticians who have only recently forayed into the realm of genetics. There were also statistical researchers present who had specialized more in the area of matching of DNA sequences while others have specialized in the localization of disease genes.

The first day of the workshop was devoted to DNA sequence matching and issues related to forensics. The second and third days focussed on problems of modeling phylogenies and inferential difficulties related to the complex tree structures produced, as well as the method of coalecence. The fourth and fifth days centered around developments related to identifying disease genes.

The field of statistical genetics is growing and expanding. Though the Genome Project will eventually result in the sequencing of the human genome, as well as the genomes of several other organisms, there will still be a need for good statistics for family studies of complex diseases. Simple sequencing of the genome will not necessarily provide insight to how a disease or susceptibility to environmental influences is regulated by a complex interaction of several genes. Of special interest is the growing recognition of the potential role of interaction of mitochondrial genes with nuclear genes to produce many chronic or degenerative disorders. Statistical genetics and the development of appropriate study designs are still in the very early stages for this aspect of human genetics.

The use of phylogenies and coalescences is widespread in evolutionary biology, in studies of population flow or establishment of relatives when applied to mitochondrial DNA, as well as in establishing routes of infections, as the case of human immunodeficiency syndrome. There is still much room for improving model building and particularly in understanding inference in this arena.

The use of statistics for assessing identification in criminal and paternity cases through mitochondrial of nuclear DNA is also becoming more widespread. The uncertainty and controversy over these methods is likely to rage for many years to come.

The papers in this volume are a sample of the topics that were discussed during the workshop. We thank those participants who were willing to contribute to this record of a lively and fruitful meeting.

M. Elizabeth Halloran, Emory University

Seymour Geisser, University of Minnesota

July 1998

## CONTENTS

# THE DNA TYPING CONTROVERSY AND NRC II

LAURENCE D. MUELLER*

The application of modern molecular biological techniques of DNA (deoxyribonucleic acid) typing to forensic science has been a major advance. These techniques aid in the identification of people who may be the sources of biological tissue samples left at the scene of crimes. The application of these techniques in forensic science has not been without controversy. One consequence of this controversy has been the creation of two different committees by the NRC (National Research Council), a branch of the National Academy of Sciences, to review these problems. The first committee issued their report in April of 1992 (NRC, 1992) and suggested some major changes in the methods used to report statistics by forensic DNA typing laboratories. Four years later another NRC committee issued a second report (NRC, 1996) which suggested that many of the recommendations of the first NRC committee were unnecessary. In this paper I review some of the population genetic and statistical issues which have been at the heart of the debate in forensic DNA typing. I also analyze, for the first time, forensic PCR (polymerase chain reaction) databases from five different laboratories. Finally, I discuss some of the specific recommendations made by the second NRC committee (also called NRC II).

**1. Why there is a controversy with forensic DNA typing.** There are two scientific disciplines whose techniques and theories are used in the development of DNA typing technology today: molecular genetics and population genetics. For restriction fragment length polymorphism (RFLP) based techniques the theories and principles of molecular genetics are used for developing the techniques to isolate DNA from evidence samples, to break up the DNA into small fragments and then to finally visualize these fragments on x-ray films or autorads. Once these patterns are visible and it has been determined that DNA from an evidence sample and DNA from a suspect "match" the theories and principles of population genetics are used to estimate how rare people with matching profiles might be in the potential suspect population. This statistic is often called the match probability (or genotype frequency) and is most often computed by a method known as the product rule.

With DNA typing techniques one can easily get predicted frequencies of one in millions to one in trillions as estimates of match probabilities. If these procedures yielded match probabilities of 1 in 100, say, then a moderate number of people with matching profiles would be expected in a database of several thousand people. An excess or deficiency of people matching the profile would serve as empirical refutation of the product rule.

*Department of Ecology and Evolutionary Biology, University of California, Irvine, CA 92697-2525. Email: ldmuelle@uci.edu.

This type of independent check is not possible if most genetic profiles are not present in the existing data bases. Consequently, there is a far greater need to verify that the techniques which generate these very rare predicted frequencies are accurate. Certainly, the estimated frequencies should not exaggerate the rarity of the genetic pattern.

**2. Investigating the assumptions and reliability of the product rule.** A DNA profile is composed of genetic information from several different sites on the hereditary material, called DNA. These locations are referred to as genes, loci or (incorrectly) probes. There are two copies of each gene, a maternally derived copy and a paternally derived copy. When there exists, within a *population*, different forms of a gene the different forms are called alleles. As an example for a particular gene, say $A$, if there exist ten different forms, they can be represented by the symbols $A_1, A_2, ..., A_{10}$. An individual may be either a heterozygote at the $A$ locus, meaning the individual has two different copies of the gene (e.g. $A_2A_6$) or the individual may be a homozygote, meaning the individual has two copies of the same allele (e.g. $A_7A_7$). A profile usually consists of several loci, say $A, B, C$ and $D$, thus a single individuals profile may be represented as, $A_iA_jB_iB_jC_iC_jD_iD_j$.

The first step in the product rule is to estimate the frequency, within an appropriate reference population, of the combination of alleles at a single locus (e.g. the frequency of people having the $C_iC_j$ combination of alleles). This is accomplished by a relationship known in population genetic theory as the Hardy-Weinberg law. This law states that for heterozygotes the frequency of people with that particular combination of alleles is twice the product of the constituent allele frequencies (thus if the $C_i$ allele occurs with frequency 0.1 in the reference population and the $C_j$ allele with frequency 0.2 then people with the $C_iC_j$ combination should occur with frequency $2 \times 0.1 \times 0.2 = 0.04$). If the individual is a homozygote the frequency of this pattern is given by the square of the individuals allele frequency (thus if the $C_i$ allele occurs with frequency 0.1 in the reference population then people with the $C_iC_i$ combination should occur with frequency $0.1 \times 0.1 = 0.01$).

The next step in the product rule is the multiplication of each of the Hardy-Weinberg results from each locus by each other. This multiplication is only permissible if the population of interest obeys the condition of linkage equilibrium (thus if the Hardy-Weinberg frequencies at the $A, B, C$ and $D$ loci were $0.1, 0.15, 0.2$ and $0.25$ respectively, the final profile frequency would be, $0.1 \times 0.15 \times 0.2 \times 0.25 = 0.00075$).

Neither Hardy-Weinberg nor linkage equilibrium are guaranteed to hold from first principles. On the contrary many biological phenomenon can invalidate these assumptions (Hartl and Clark, 1989, pgs. 31–32; Lewontin and Hartl, 1991). The biological phenomenon that poses the most serious threat to the use of the product rule is the possibility that the reference populations (usually African Americans, Caucasians and Hispanics) are composed of population subgroups that are genetically different from each other (e.g. if Hispanic databases were samples from Cuban Hispanics and Mexican Hispanics and these were genetically different from each other). Population subgroups typically appear because mating between members of these groups has been limited for some period in the past. Consequently, there have been several attempts to investigate the accuracy of the product rule by testing its key assumptions of Hardy-Weinberg and linkage equilibrium.

**2.1. The hypothesis testing paradigm.** The appropriateness of assuming Hardy-Weinberg and linkage equilibrium may be investigated by statistical hypothesis tests. Some tests of the Hardy-Weinberg equilibrium simply compared expected and obeserved frequencies of total homozygotes and heterozygotes (Odelberg et al., 1989; Budowle et al., 1991; Weir, 1992a). When used with RFLP databases these tests generally suggested an excess of homozygotes in the samples. Devlin et al. (1990) developed a goodness of fit test of Hardy-Weinberg which was designed to take into account one artifact of the RFLP technique which could generate the observed excess of homozygotes. Geisser and Johnson (1993) use a method based on quantiles and have found evidence of departures from Hardy-Weinberg equilibrium in the FBI databases. Weir (1992a) developed likelihood ratio tests for independence within and between loci. Weir (1992a,b) typically found departures from independence when the entire database was analyzed but few departures when the analysis was restricted to heterozygotes. This latter technique was justified because some heterozygotes may appear as homozygotes due to technical limitations of the RFLP process. Many of the tests described above can not be easily extended to multiple loci because of the limited size of the existing databases and the large number of multilocus genotypes that are possible with these genes.

However, there are some tests which permit the detection of multilocus dependence even with small databases. Detecting independence at three or more loci is important since it is theoretically possible that all pairs of loci may be in linkage equilibrium but the combination of three or more may not be (Feldman *et al.*, 1974). The number of heterozygous loci for each individual forms the basic observation for the test developed by Brown and Feldman (1980). This test, while easy to apply, is not very powerful especially since there are many configurations of linkage disequilibrium that will not be detected by this technique. More recently, there has been work on applying Fisher's exact tests to genetic systems with many alleles and many loci (Guo and Thompson, 1991; Maiste and Weir, 1995; Zaykin, Zhivotovsky and Weir, 1995). These tests work by computing the probability of the observed sample, under the hypothesis of independence, to randomly shuffled databases. The frequency of shuffled databases that are less likely to occur than the observed database constitute the $p$-value of the test.

**2.1.1. Limitations of the hypothesis testing paradigm.** The value of the hypothesis testing results are limited for several different reasons. If the product rule is used with four loci, say, then one must show that there is simultaneous linkage equilibrium for all four loci. Presently only the exact tests and the Brown and Feldman tests can be used with more than two loci for most existing databases. At the time the second NRC report was issued there had been no application of the exact tests to two or more loci and only a few instances of the less powerful Brown and Feldman test (Budowle et al., 1995). Interestingly, in the Budowle et al. (1995) paper the exact test is used for within locus tests of independence but not for tests across multiple loci. The only test used by Budowle et al. for more than two loci was the Brown and Feldman test. For some databases (especially very small ones) and for some statistical tests the product rule could be false but the test would fail to detect this due to low power. This issue will be discussed in more detail in the next section. With RFLP testing techniques it is possible for certain heterozygotes to appear as homozygotes (Devlin et al., 1990; Mueller, 1991). This makes testing for independence more difficult (Zaykin et al., 1995).

Another issue which needs to be considered is the way to handle datasets which have been subjected to a large number of multiple tests of independence. Several studies have sought to protect the level of type I errors, even with as few as nine different tests, with simultaneous tests statistics such as the Bonferroni inequality (Devlin et al., 1990; Budowle et al., 1995). There has been little discussion in these publications about the power of the resulting tests (see next section), although it is clear that one ultimately sacrifices statistical power by utilizing simultaneous test statistics (Miller, 1966).

**2.1.2. Applications of hypothesis tests to forensic RFLP databases.** There have been many studies now attempting to test Hardy-Weinberg, linkage-equilibrium or both in a variety of forensic and non-forensic databases (Odelberg *et al.*, 1989; Devlin *et al.*, 1990; Budowle *et al.*, 1991; Chakraborty and Daiger, 1991; Devlin and Risch, 1992; Risch and Devlin, 1992; Edwards *et al.*, 1992; Weir, 1992a, 1992b; Geisser and Johnson, 1993, 1995; Slimowitz and Cohen, 1993; Maiste and Weir, 1995). The results of these studies have been mixed. Some studies have yielded results inconsistent with either Hardy-Weinberg or linkage equilibrium (Odelberg et al., 1989; Budowle et al., 1991; Edwards et al., 1992; Weir, 1992a; Geisser and Johnson, 1993, 1995; Slimowitz and Cohen, 1993; Maiste and Weir, 1995), while others have generally supported these assumptions (Devlin *et al.*, 1990; Chakraborty and Daiger, 1991; Devlin and Risch, 1992; Risch and Devlin, 1992; Weir, 1992a, 1992b).

Some of these tests suffer from a clear lack of power. For instance Devlin *et al.* (1990) were the first to develop tests specifically for these RFLP data sets and claim their results appeared to conform to the Hardy-

Weinberg expectations. Devlin et al. examined three databases (Lifecodes) and three probes for a total of nine different tests of the Hardy-Weinberg expectations. These tests produced a number, $z$, which measured the departure from Hardy-Weinberg. If this number was greater than -2.8 and less than 2.8 the results were judged consistent with Hardy-Weinberg. Devlin et al.'s test statistic incorporated the simultaneous test protection of the Bonferroni inequality. The problem with using the Bonferroni inequality is the potential to reduce the statistical power of the tests.

This problem was raised by several commentators on Devlin *et al.'s.* work (Cohen, *et al.*, 1991; Green and Lander, 1991). In response to these criticisms Devlin *et al.* (1991) conducted numerical studies which seemed to suggest that the original test was quite powerful. However the numerical studies by Devlin *et al*'s. (1991) used a different statistical test than in their original paper. The new test both removed the Bonferroni protection and changed the original two-tailed test to a one-tailed test. With this altered test $z$ was now required to be greater than $-1.65$ and less than 1.65 to be considered consistent with Hardy-Weinberg. If this version of their test is applied to their original results, the Lifecodes databases show two significant departures from the Hardy-Weinberg expectations.

More recently Zaykin et al. (1995) have analyzed the power of the exact test to detect departures from independence caused by population substructure. The power of these tests was reasonable when the coancestry coefficient, $\theta$, (which measures the extent of population substructure and which the second NRC committee suggested would typically be between 0.01 and 0.03 in most U. S. populations) was in the range of 0.05 to 0.1. In fact the power gets better with increasing numbers of loci. However, when the exact test is applied to the FBI's VNTR database there are numerous examples of departures from Hardy-Weinberg and linkage equilibrium (two-loci, Maiste and Weir, 1995). Maiste and Weir attribute these departures to the existence of small missing bands which cause some heterozygotes to be misclassified as homozygotes. They thus reanalyzed these data, examining only the heterozygotes and omiting the homozygotes. Their conclusion was that this type of analysis showed that, ".. there is overall evidence for independence." However, this conclusion stands in marked contrast to the conclusions Zaykin *et al.* (1995). In that paper the power of the exact test with homozygotes excluded was barely at the nominal level of 5% for tests with two-loci and $\theta$ at 0.05 (see table 4, Zaykin *et al.*, 1995). The power of these tests actually gets worse with increasing number of loci, eventually having power less the nominal level of 5%.

**2.1.3. Application of hypothesis tests to forensic DQ-A and polymarker databases.** In addition to the RFLP DNA tests are a variety of DNA tests which utilize the polymerase chain reaction (PCR). This technique permits the amplification of very small amounts of DNA in an evidence sample until there is enough to be tested by a variety of techniques.

The net result is that samples too small to provide results by RFLP techniques may be analyzed by PCR techniques. The DQ-A gene and five genes known as the polymarker genes are amplified by PCR techniques and used in forensic DNA typing.

Budowle et al. (1995) analyzed the FBI's African American, Caucasian and Hispanic databases for independence and concluded there was general agreement with the assumptions of the product rule. I present results below which question this conclusion. My own analysis of just the FBI database has recently turned up one error in the results published by the FBI (Mueller, 1998). It appears that application of the exact test to the FBI's Caucasian database and the HBGG locus yields a highly significant departure from equilibrium ($p = 0.008$) which was reported by Budowle et al. as non-significant ($p = 0.887$). Budowle et al.'s result can only be obtained if one changes the genotype of a $CC$ homozygote (the only individual with a $C$-allele in the FBI database) to an $AA$ homozygote.

A major limitation of the FBI database for assessing independence is its relatively small size, with samples sizes between 94 and 148 people. This shortcoming can be remedied by pooling data from several other forensic laboratories which have analyzed the same loci and presumably the same populations. I have done this for databases obtained from the FBI, Cellmark, the Minnesota Bureau of Criminal Apprehension, Perkin-Elmer and the Virginia State Crime Laboratory. All five databases include samples from Caucasians and African American's. Only three laboratories had Hispanic databases, the FBI, Perkin-Elmer and the Virginia State Crime Lab. For this analysis the Florida and Texas FBI Hispanic databases were combined. All test results for independence within loci are shown in Tables 1-3 along with the significant departures from multilocus equilibrium.

I have not set as a pre-condition for these tests a comparison of the allele frequencies in the pooled databases. Thus, its possible that genetic differences that exist among the sampled populations contribute to the lack of independence. However, one must remember that the pooling of these samples is in fact no different than the sampling process that originally gave rise to these databases. Since there has been no attempt to insure these sample are random it is important to also determine if ad hoc samples pooled together obey the independence laws. If they do not, then all the current databases would need to be recreated with closer attention paid to the sampling process.

There are only two significant departures from Hardy-Weinberg equilibrium, but they are both in the Caucasian population (HBGG and GC, Table 2). There are a total of 57 different multilocus tests that were performed on each database. Since there were two different tests done (Fisher's and the chi-square exact, see Zaykin et al., 1995) there are actually a total of 114 tests per population. Assuming all tests are independent (which of course they are not) we might expect about 5 or 6 significant results. For the African American population (Table 1) there were 15 significant departures (including the combined polymarker and DQ-A loci) while for the Caucasian population (Table 2) there were 11. However, for the Hispanic population (Table 3) there were only two significant results. It would appear that both the Caucasian and African American population are yielding evidence of significant departures from independence while the Hispanic population is not. At this point it is also worth noting that the Hispanic database is about half the size of the other two databases and hence we expect the tests of independence in the Hispanic populations to be less powerful.

TABLE 1

*Summary of results from exact tests of independence for the combined African America dataset. All single locus results are shown. Only those multilocus tests with at least one significant ($p < 0.050$) departure from equilibrium are shown. There were a total of 57 different multilocus tests performed. Estimated p-values are based on 3200 permutations of the database, thus an approximate confidence interval on a p-value of 0.05 is about ±0.0076. For the tests within loci the permutations broke up allele combinations at a single locus. For the tests between loci allele combinations at a single locus were preserved and genotypes between loci were randomly shuffled. This latter test procedure appears to the most powerful for detecting departures from linkage equilibrium and will not be influenced by departures from Hardy-Weinberg at the constituent loci (Zaykin et al., 1995).*

| Loci | N | Fisher's Test, $p$ | Chi-Square Test, $p$ |
|---|---|---|---|
| DQ-$\alpha$ | 638 | 0.23 | 0.23 |
| LDLR | 638 | 0.31 | 0.27 |
| GYPA | 638 | 0.88 | 0.93 |
| HBGG | 638 | 0.78 | 0.81 |
| D7S8 | 638 | 0.31 | 0.28 |
| GC | 638 | 0.28 | 0.28 |
| Loci Combination | | | |
| DQ-$\alpha$/HBGG/GC | | 0.0041 | 0.013 |
| LDLR/HBGG/GC | | 0.0088 | 0.053 |
| DQ-$\alpha$/LDLR/HBGG/GC | | 0.033 | 0.0053 |
| DQ-$\alpha$/GYPA/HBGG/GC | | 0.089 | 0.014 |
| DQ-$\alpha$/HBGG/D7S8/GC | | 0.063 | 0.043 |
| LDLR/GYPA/HBGG/GC | | 0.0052 | 0.011 |
| LDLR/HBGG/D7S8/GC | | 0.028 | 0.056 |
| DQ-$\alpha$/LDLR/GYPA/HBGG/GC | | 0.30 | 0.0053 |
| DQ-$\alpha$/LDLR/HBGG/D7S8/GC | | 0.079 | 0.0094 |
| DQ-$\alpha$/GYPA/HBGG/D7S8/GC | | 0.22 | 0.048 |
| LDLR/GYPA/HBGG/D7S8/GC | | 0.18 | 0.011 |
| DQ-$\alpha$/LDLR/GYPA/HBGG/D7S8/GC | | 0.24 | 0.011 |

As mentioned previously it has been suggested that techniques, like the Bonferroni inequality, be used in the situations of multiple testing to insure a type-I error of no greater than 5%. My own feeling is that this

TABLE 2

*Summary of results from exact tests of independence for the combined Caucasian dataset. All single locus results are shown. Only those multilocus tests with at least one significant ($p < 0.050$) departure from equilibrium are shown. There were a total of 57 different multilocus tests performed. The test procedure follows the methods in table 1.*

| Loci | N | Fisher's Test, $p$ | Chi-Square Test, $p$ |
|---|---|---|---|
| DQ-$\alpha$ | 624 | 0.93 | 0.93 |
| LDLR | 624 | 0.28 | 0.30 |
| GYPA | 624 | 0.82 | 0.88 |
| HBGG | 624 | 0.0028 | 0.0013 |
| D7S8 | 624 | 0.73 | 0.68 |
| GC | 624 | 0.010 | 0.010 |
| Loci Combination | | | |
| LDLR/HBGG | | 0.049 | 0.51 |
| HBGG/GC | | 0.20 | 0.025 |
| DQ-$\alpha$/LDLR/HBGG | | 0.0091 | 0.61 |
| DQ-$\alpha$/HBGG/D7S8 | | 0.021 | 0.53 |
| LDLR/D7S8/GC | | 0.45 | 0.044 |
| GYPA/HBGG/GC | | 0.056 | 0.013 |
| HBGG/D7S8/GC | | 0.39 | 0.034 |
| DQ-$\alpha$/LDLR/HBGG/D7S8 | | 0.029 | 0.56 |
| GYPA/HBGG/D7S8/GC | | 0.18 | 0.025 |
| LDLR/GYPA/HBGG/GC | | 0.088 | 0.035 |
| LDLR/GYPA/HBGG/D7S8/GC | | 0.14 | 0.049 |

TABLE 3

*Summary of results from exact tests of independence for the combined Hispanic dataset. All single locus results are shown. Only those multilocus tests with at least one significant ($p < 0.050$) departure from equilibrium are shown. There were a total of 57 different multilocus tests performed. The test procedure follows the methods in table 1.*

| Loci | N | Fisher's Test, $p$ | Chi-Square Test, $p$ |
|---|---|---|---|
| DQ-$\alpha$ | 381 | 0.090 | 0.12 |
| LDLR | 381 | 0.92 | 0.92 |
| GYPA | 381 | 0.13 | 0.14 |
| HBGG | 381 | 0.71 | 0.76 |
| D7S8 | 381 | 0.74 | 0.74 |
| GC | 381 | 0.90 | 0.87 |
| Loci Combination | | | |
| LDLR/GYPA | | 0.023 | 0.033 |

sacrifice in power can't be justified. For instance if the Bonferroni inequality were used on the 57 multilocus tests in Tables 1-3 the $p$-value needed for significance would be 0.0009. To assess the implications of such a procedure I have performed Fisher's exact test on three biallelic loci from the FBI's Caucasian database, LDLR, D7S8 and GYPA. However, in these tests I actually saved the lowest 5% of the shuffled database based on their probability of being observed and I saved the lowest 0.09% (out of 10,000 permutations). These extreme databases all showed a deficiency of heterozygotes. The magnitude of this deficiency can be quantified with the parameter $\theta$ from the following relationship,

$$[\text{extreme heterozygote frequency}] = (1-\theta)[\text{Hardy-Weinberg frequency}]$$

The results (Table 4) show that in general adding the Bonferroni protection requires that the magnitude of the departures from Hardy-Weinberg be almost twice as large as tests without the Bonferroni protection.

TABLE 4

*The magnitude of population substructure, as measured by $\theta$ that is necessary to cause rejection of the Hardy-Weinberg null hypothesis via Fisher's exact test. The databases used are from the FBI's Caucasian polymarker database.*

| | $\theta$ value necessary for | |
|---|---|---|
| Locus | Rejection at the 5% level | Rejection at the 0.09% level |
| D7S8 | 0.17 | 0.26 |
| LDLR | 0.15 | 0.26 |
| GYPA | 0.15 | 0.26 |

Another class of genetic markers which can be amplified by PCR techniques are called short tandem repeats (STR's). Several forensic laboratories have started to use these genetic markers in case work. In Tables 5-6, I show tests of independence for the Cellmark African American (Table 5) and Caucasian (Table 6) databases. These tests include DQ-$\alpha$, polymarker and the three STR loci HUMCSF1PO, HUMTPOX, and HUMTHO1 for nine loci altogether. Each of these databases is rather small (100-103 individuals) and there were a total of 502 multilocus tests done per population. The African American database yielded 21 significant departures which is in the ball park of the nominal number expected if all tests are considered independent. The Caucasian database yielded more than twice as many significant results (49), including the combinations of $HUMCSF1PO$ and $HUMTPO1$ and the non-STR loci. Taken altogether, the results in Tables 1-6 suggest that for many PCR based systems there appears to be evidence of departures from statistical independence.

**2.2. Investigation of population subgroups.** Another methodology to test the suitability of the product rule is to investigate the underlying assumptions of Hardy-Weinberg and linkage equilibrium. An important underlying assumption is that the populations utilized are homogeneous, randomly mating populations. They should not be composed of genetically differentiated subgroups. In an attempt to study this problem the FBI has collected data from laboratories throughout the United States and from many other countries (Budowle *et al.*, 1994a, b). Many of these databases

TABLE 5

*Exact test of independence for the African American Cellmark database (100 people). These tests were evaluated by using the chi-square statistic (Zaykin, Zhivotovsky and Weir, 1995). The test procedures were in Table 1. The loci used are numerically identified as follows.* **1** *- LDLR,* **2**- *GYPA,* **3** *- HBGG,* **4**- *D7S8,* **5** *- GC,* **6** *- DQ-a,***7** *- HUMCSF1P0,* **8** *- HUMTPOX,* **9** *- HUMTHO1.*

| Probability ($p$-value) | Loci (locus) |
|---|---|
| 0.013 | 1 |
| 0.045 | 1/5/7 |
| 0.036 | 1/6/7 |
| 0.021 | 2/3/8 |
| 0.013 | 2/3/9 |
| 0.048 | 3/6/9 |
| 0.0084 | 5/6/7 |
| 0.048 | 1/2/5/7 |
| 0.039 | 1/2/6/7 |
| 0.0041 | 1/5/6/7 |
| 0.035 | 2/3/4/9 |
| 0.046 | 2/3/6/9 |
| 0.0066 | 2/5/6/7 |
| 0.015 | 3/5/6/7 |
| 0.043 | 4/5/6/7 |
| 0.0034 | 1/2/5/6/7 |
| 0.013 | 1/3/5/6/7 |
| 0.025 | 1/4/5/6/7 |
| 0.020 | 2/3/5/6/7 |
| 0.014 | 1/2/3/5/6/7 |
| 0.027 | 1/2/4/5/6/7 |

are not samples of population subgroups, rather they are simply samples from heterogeneous groups (like Caucasians) from different geographic regions of the United States.

TABLE 6

*Exact tests of independence for the Caucasian Cellmark database (103 people). These tests were evaluated by using the chi-square statistic (Zaykin, Zhivotovsky and Weir, 1995). The loci used are numerically identified as follows.* **1**-*LDLR,* **2** *- GYPA,* **3** *- HBGG,* **4** *- D7S8,* **5** *- GC,* **6** *- DQ - a,* **7** *- HUMCSF1PO,* **8** *- HUMTPOX,* **9** *- HUMTHO1.*

| Probability ($p$-value) | Loci (locus) |
|---|---|
| 0.029 | 5 |
| 0.041 | 9 |
| 0.0044 | 1/3 |
| 0.033 | 2/6 |
| 0.038 | 1/3/8 |
| 0.036 | 1/5/7 |
| 0.033 | 2/3/8 |
| 0.046 | 2/5/7 |

| | |
|---|---|
| 0.049 | 3/4/8 |
| 0.033 | 3/5/8 |
| 0.0094 | 3/6/8 |
| 0.041 | 1/2/3/8 |
| 0.013 | 1/2/5/7 |
| 0.035 | 1/3/5/8 |
| 0.0097 | 1/3/6/8 |
| 0.036 | 2/3/5/8 |
| 0.0084 | 2/3/6/8 |
| 0.044 | 3/4/5/8 |
| 0.0097 | 3/4/6/8 |
| 0.0088 | 3/5/6/8 |
| 0.038 | 3/6/8/9 |
| 0.040 | 1/2/3/5/8 |
| 0.011 | 1/2/3/6/8 |
| 0.041 | 1/2/4/5/7 |
| 0.043 | 1/3/4/5/8 |
| 0.012 | 1/3/4/6/8 |
| 0.0084 | 1/3/5/6/8 |
| 0.048 | 1/3/6/7/8 |
| 0.042 | 1/3/6/8/9 |
| 0.044 | 2/3/4/5/8 |
| 0.011 | 2/3/4/6/8 |
| 0.012 | 2/3/5/6/8 |
| 0.050 | 2/3/6/7/8 |
| 0.040 | 2/3/6/8/9 |
| 0.014 | 3/4/5/6/8 |
| 0.045 | 3/5/6/7/8 |
| 0.046 | 3/5/6/8/9 |
| 0.046 | 1/2/3/4/5/8 |
| 0.013 | 1/2/3/4/6/8 |
| 0.012 | 1/2/3/5/6/8 |
| 0.047 | 1/2/3/6/7/8 |
| 0.039 | 1/2/3/6/8/9 |
| 0.012 | 1/3/4/5/6/8 |
| 0.048 | 1/3/5/6/7/8 |
| 0.015 | 2/3/4/5/6/8 |
| 0.041 | 2/3/5/6/8/9 |
| 0.015 | 1/2/3/4/5/6/8 |
| 0.043 | 1/2/3/4/6/8/9 |
| 0.045 | 1/2/3/5/6/7/8 |
| 0.041 | 1/2/3/5/6/8/9 |
| 0.050 | 1/2/3/4/5/6/8/9 |

However, some of the samples do come from different population subgroups, e.g. Swiss Caucasians, and East Indians and can be used to directly assess the assumption of no genetic differentiation. This is done most directly by simply comparing the frequency of the different forms of these genes in two populations. Figure 1 shows one such example from the FBI's world wide study.

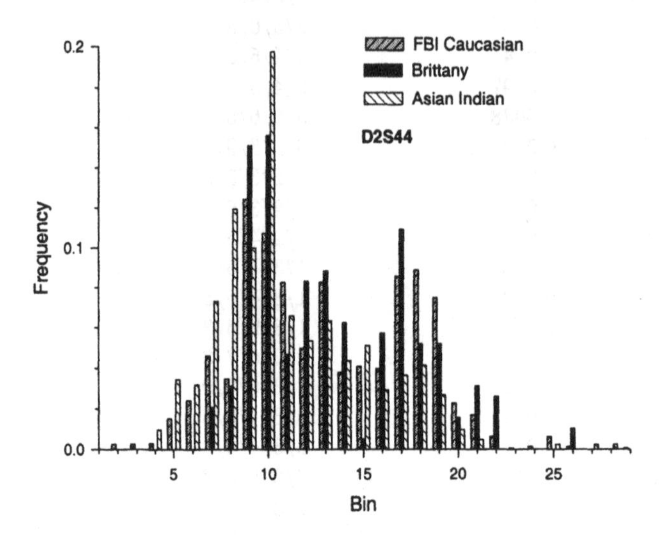


FIG. 1. *The frequency of the FBI fixed bins in three different populations. The differences that exist in these three populations is statistically significant meaning it is too great to be due to chance events associated with the sampling process itself.*

This example is only one of many. Recently, Sawyer *et al.* (1996) reported that about 70% of all such comparisons in the Caucasian populations in the world wide study show significant differences. Data from other studies (Balzas *et al.*, 1992; Deka *et al.*, 1991; Krane *et al.*, 1992; Sanjantila *et al.*, 1992) have also confirmed the general conclusion that many population subgroups are significantly differentiated for these VNTR genes.

I have recently completed a similar analysis as Krane *et al.*, with the Hinfl databases and the U.S. Caucasian database of Cellmark. Many populations subgroups are not represented in this study. For instance there were no subgroups of Hispanics studied e.g. Cubans, Puerto Ricans, Haitians, Mexican Americans, etc. Likewise the study of Asian Americans was very superficial consisting of two Chinese samples neither with more than 20 people, a number wholly inadequate for any reasonable inferences. However, the world wide study does contain some reasonable data on Caucasians which I have analyzed.

There are several different ways one might analyze the significance of population subgroups. One could, in principle compare the frequencies of a number of five locus profiles to see if they differ from one population to the next. However, this is practically impossible since the Cellmark databases contain very few people which have been typed at all five loci used by Cellmark. For instance in Cellmark's African American population there are only two people with complete five locus profiles. The FBI has studied subgroups by estimating five locus profile frequencies by using the product rule. This is not a reliable procedure since there is no way to independently verify the accuracy of product rule frequencies.

The most direct and reliable way to assess whether there are genetic differences between population subgroups is to compare the observed frequency of DNA fragments of different sizes. I have used this procedure with the Cellmark Caucasian database and a variety of other population groups (Table 7). The FBI fixed bins were used as a standard so that different populations could be compared to each other. There is no easy way to do this comparison using the floating bin procedure. For each pair of populations a likelihood ratio test was used to assess whether the fixed bin frequencies were significantly different. If the difference is very unlikely to be due to chance alone it is then reasonable to conclude that the genetic differences between the two populations are real.

In Table 7, 32 of 49 comparisons show significant differences or 65% of all contrasts. Thus, these data support the existence of widespread population substructure. Consequently, one needs to be concerned about the lack of random sampling on the part of Cellmark since these data suggest that the frequencies of genetic variants detected by the RFLP techniques of Cellmark do vary significantly from one population to the next. In some cases these differences are large. In figure 2, I graphically show some of these differences.

While these studies confirm the widespread existence of allele frequency differences among population subgroups the more difficult question is whether there are differences among the multilocus profiles. While it seems improbable that there can be allele frequency differences and not multilocus profile differences determining the magnitude of this effect is difficult.

One approach taken by the FBI is to use the product rule to compute the frequency of a single profile in multiple databases. Observing that this procedure typically gave different, but rather rare frequencies the FBI has concluded (Budowle et al., 1994 a,b) that population subgroups lead to no "forensically" significant differences among the profile frequencies. While the phrase "forensically significant" sounds like "statistically significant" the basis of the concepts is quite different. Statistical significance has an objective meaning which can be easily understood while forensic significance might mean something different to every person. In principle two frequency estimates would be considered forensically significant if a juror would render a different verdict depending on which number they heard. Of course determining when differences are forensically significant can only

TABLE 7

*Results from a likelihood ratio test comparing the bin frequencies in the Cellmark Caucasian data base and one of the data bases below. The first number in each cell is the G statistic, the subscript is the degrees of freedom and the probability of observing this statistic is given in parentheses. The differences between the two populations are judged significant when the probability of observing the test results is 5% or less. The significant results are listed in boldface type.*

| Population | D1S7 | D2S44 | D7S21 | D7S22 | D12S11 |
|---|---|---|---|---|---|
| Alsatian | | $29_{17}$ **(0.031)** | | | $24_{13}$ **(0.033)** |
| Danish | $42_{24}$ **(0.011)** | $31_{18}$ **(0.028)** | $21_{14}$ (0.11) | | $31_{13}$ **(0.004)** |
| English | $52_{25}$ **(0.001)** | $32_{18}$ **(0.022)** | $24_{18}$ (0.17) | $131_{25}$ **(<0.001)** | $17_{14}$ (0.26) |
| Finnish | | $29_{14}$ **(0.01)** | | | |
| German | $60_{27}$ **(<0.001)** | $50_{17}$ **(<0.001)** | $19_{15}$ (0.22) | $34_{24}$ (0.09) | $51_{14}$ **(<0.001)** |
| Italian | | $33_{18}$ **(0.018)** | | | $42_{14}$ **(<0.001)** |
| New Zealander | | $22_{15}$ (0.12) | | | $17_{12}$ (0.14) |
| Norwegian | | $24_{17}$ (0.11) | $19_{12}$ (0.092) | $73_{24}$ **(0.028)** | $37_{12}$ **(<0.001)** |
| South Europe | $36_{22}$ **(0.031)** | | $20_{13}$ (0.087) | | $36_{13}$ **(0.001)** |
| Spanish | | $25_{18}$ (0.12) | $18_{13}$ (0.15) | | $46_{13}$ **(<0.001)** |
| Swiss | | $46_{18}$ **(<0.001)** | $19_{16}$ (0.26) | $46_{24}$ **(0.004)** | $76_{15}$ **(<0.001)** |
| Swedish | $37_{25}$ (0.055) | $24_{18}$ (0.15) | $17_{12}$ (0.14) | $31_{24}$ (0.16) | $52_{12}$ **(<0.001)** |
| Asian Indian | $67_{27}$ **(<0.001)** | $47_{16}$ **(<0.001)** | $36_{19}$ **(0.01)** | $151_{26}$ **(<0.001)** | $132_{14}$ **(<0.001)** |
| Maori | | $102_{14}$ **(<0.001)** | | | $207_{12}$ **(<0.001)** |
| Pacific Islander | | $184_{15}$ **(<0.001)** | | | $372_{13}$ **(<0.001)** |

be done if the scientist can determine and quantify how "jurors" evaluate and weigh evidence. There is no agreed upon process for making these conclusions (Koehler *et al.*, 1995).

As discussed earlier the preferred method for evaluating the accuracy of the product rule would be to compare the predicted frequencies to the observed. In some cases this can be done. The profiles in table 8 are for two individuals, one from the Maya population in Mexico and the second from the Surui population in Brazil (Kidd *et al.*, 1991).

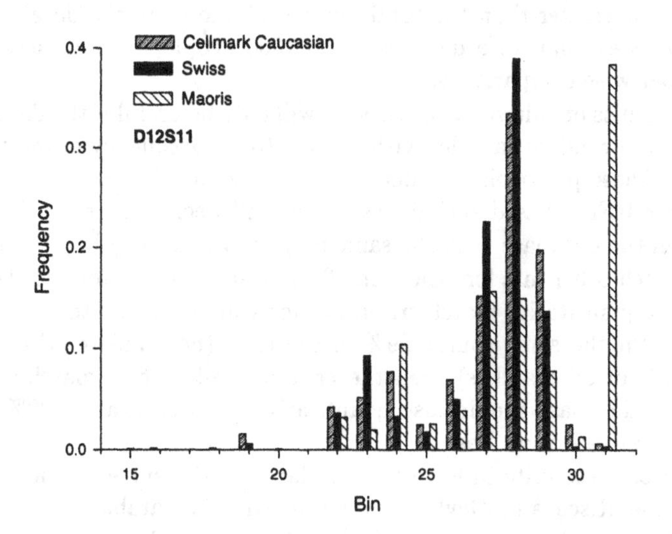


FIG. 2. *The frequency of fixed bins in three populations for the D12S11 locus. The frequencies of these fixed bins in the Maoris and Swiss populations are significantly different than the Cellmark Caucasian database.*

TABLE 8

*The size (in base pairs) of each allele at six loci for two unrelated individuals from the Surui and Maya populations.*

| Individual | D2S44 | D17S79 | D14S1 | D14S13 | D18S27 | LILA5 |
|---|---|---|---|---|---|---|
| Surui | 10,600 | 3,570 | 4,570 | 12,790 | 4,750 | 7,330 |
| | 10,290 | 3,320 | 3,790 | 5,210 | 4,520 | 7,330 |
| Maya | 10,820 | 3,560 | 4,550 | 12,780 | 4,750 | 7,320 |
| | 10,270 | 3,290 | 3,790 | 5,180 | 4,550 | 7,320 |

The product rule predicts that the profile the two individuals in table 8 share in common should have a frequency of 1 in 96 million, although the observed frequency is 1 in 37. This large discrepancy stands in contrast to the FBI and NRC II's assertions that subgroups typically alter profile frequencies by only a factor of 10-100.

**2.4. How rare are matches between unrelated people?** Another research protocol is to determine how common matching profiles are in forensic databases (Risch and Devlin, 1992). This is done by comparing the profile of each person to every other person in the database using the laboratories quantitative match criteria. For example, if there are 1000 people in the database, person number 1 is compared to person number 2 and then to person number 3 etc., then person number 2 is compared to person number 3 etc. In this way the total number of pairwise comparisons can be much greater than the total number of people in the database. In the previous example of a database with 1000 people there are a total of 499,500 pairwise comparisons.

The results of pairwise comparisons will only be useful if the databases examined are random samples with respect to the frequency of multilocus matches. These protocols also depend critically on using a correct match criteria for RFLP based techniques which will declare matches between samples in the database with the same frequency as if they had been compared to each other in a forensic test. This means that if a forensic laboratory uses a quantitative match criteria which will declare matches between samples from the same source 99% of the time (no realistic criteria will catch 100% of all matches) then the criteria used to find matching profiles in the database should also quantitatively match exactly 99% of all matches, no more or no less.

Both of these criteria have been violated in the current studies. For instance the Risch and Devlin study utilized FBI databases. However, before these databases were supplied to Risch and Devlin the FBI had already surveyed them for matching profiles. The FBI search had in fact found numerous matches. In some cases (data from A. Eisenberg, at the Texas College of Osteopathic Medicine) examination of the original records was able to verify that duplicates of the same persons blood had been sent to the FBI. The FBI appropriately deleted these matching profiles. In other cases there was no way to definitively determine if the matches were duplicates or not. Nevertheless, the FBI deleted these matches, since they were working under the assumption that such matches should not occur. A database from a black population in South Carolina contained 60 matching profiles that could not be accounted for (memorandum of 8/31/90 from J.J. Kearney to Mr. Hicks). The entire South Carolina black database was deleted and replaced with data from different populations. This problem is not limited to the FBI but has also happened in other forensic laboratories. The Metropolitan police Department Laboratory in St. Louis found four seven probe matches and one six probe match in their database and deleted these (memo of 6 May 1996, by Donna Becherer).

The upshot of this is these databases are biased samples. In many instances cited above matching samples have been deleted simply because of a belief that such matches shouldn't occur. **These databases can not then be used as research tools to investigate how rare multilocus matches are.**

A second more difficult issue concerns which match criteria should be used when searching databases for matching profiles. For instance in Risch and Devlin's study of the FBI's database they used a match criteria which was only half the size of the FBI's match criteria. To illustrate this problem consider the FBI's match criteria. When evidence and suspect DNA are compared on the same gel and autorad, the two samples are judged to match when the difference in their measured size is less than or equal to 5% of their average size. This match criteria should include greater than 99% of all matching profiles. Database samples are included on many different autorads and collected over a very long period of time. Consequently, two samples from the same source will tend show a greater separation in the database than they would had they been tested at the same time. This variability has been quantified (Geisser, 1996) and the database samples show about 1.8 times as much variation as samples on the same autorads. That is the standard deviation for replicated measurements made on different gels and autorads is 1.8 times larger than the standard deviation for measurements made on the same gel and autorad. This means that to find matches in a database with the same frequency as they would be found in casework requires using a match window which is 1.8 times larger than the standard match criteria. Using this criteria Geisser (unpublished) has found six four locus matches in the FBI's Caucasian database and one six locus match even after the filtering process mentioned earlier.

**3. Why the random match probability isn't a sufficient statistic for evaluating the weight of DNA evidence.** In the typical forensic case involving DNA evidence, DNA from an evidence sample is compared to DNA from a suspect. When these two samples match the inference that is usually made is that the DNA in the evidence came from the suspect. There are two different phenomena which can make this inference invalid. One phenomenon that can invalidate this inference is if the DNA came from a person other than the suspect who happens to have the same genetic profile as the suspect. Secondly, the DNA profiles in the evidence and the suspect may be different but due to a laboratory error a match has been declared.

Both of these phenomenon must be evaluated and quantified. The overall probability of an erroneous match would then equal the sum of the probability of a coincidental match between unrelated people and the probability of a laboratory error. If one of these probabilities is very rare and one very common the overall probability of an erroneous match will simply be equal to the more common probability. For example if the random match probability is 1 in 1 million but the laboratory error is 1 in 500 the chance of an erroneous match is about 1 in 500. In fact the only situation under which it would be appropriate to use 1 in 1 million to weight a DNA match is in a laboratory in which the rate of error was demonstrably less than 1 in 1 million. Not only have false matches been made in forensic laboratories but all the available evidence suggests that rates of false matches are in the range of 1 in hundreds to 1 in thousands (Mueller, 1993; Koehler, 1993; Koehler *et al.*, 1995). To only provide a jury with a very rare random match probability vastly overstates the weight which should be accorded a single DNA match (Hagerman, 1990; Lempert, 1991; Thompson, 1995).

Recent empirical research (Koehler *et al.*, 1995) has shown that these ideas are not well understood. Lay people who participated in mock trials tended to base their evaluations on the rarer of two statistics not the more common of the two as is appropriate. Thus, people who heard a very rare match probability and a very common laboratory error rate tended to convict at the same rate as people who only heard only the rare match probability. However, those who heard only the common laboratory error rate were much less likely to convict. Consequently, if the laboratory error rate and the match probability are presented separately the connection between the two numbers needs to be explained in some detail.

Some laboratories have suggested that their error rates are zero since they have made no errors in all the proficiency tests they have taken. The problem with this assertion is that in the context of match probabilities, usually, of the order 1 in millions or 1 in billions, a zero error rate means lab errors must be much less than the match probability. For a laboratory that has only performed 50 proficiency tests it is impossible to claim that the error rate is less than one in billions. In fact using widely accepted statistical methods the best that can be done for the laboratory performing all 50 tests correctly is to place an upper bound on how common errors might be (in this case that upper bound is 1 in 17 with 95% certainty). The only way to lower this bound is for the laboratory to do additional proficiency tests.

A logical question is, what if the laboratories error rate is really 1 in millions? If this were true then it would take an enormous effort for any laboratory to demonstrate their proficiency. At this point we can step back and take a larger view of all laboratories which do DNA testing. From this vantage point it is clear that false matches do not occur at the exceeding rare rate of 1 in millions or billions but are much more likely to be on the order of 1 in thousands. While it is certainly the case that some laboratories may be better than another there are no reason to believe that some labs may have error rates of 1 in 100's while others have error rates of 1 in billions for basically the same testing procedure. Consequently, laboratories should use the upper 95% confidence interval for their own individual estimate of error when they have a perfect record, or if the number of test they have done is very small the laboratory may use an estimate take from the industry in general.

**4. The second report on DNA typing from the national research council.** The first report of the National Research Council (NRC, 1992) suggested a number of changes to the standard procedures used by forensic laboratories for computing match probabilities. Many of these suggestions were aimed at resolving some of the uncertainties generated by population subgroups and some suggestions dealt directly with the proper reporting and quantification of laboratory error rates.

There were numerous complaints about several recommendations of the first NRC committee and most of the important recommendations were never implemented by forensic laboratories. In response to this situation and requests directly from agencies like the FBI a second committee was created by the National Research Council to reconsider mainly issues of statistics and population genetics. This second committee issued their report in May, 1996 (NRC, 1996). Below I review some of the important conclusions in that report.

A major reason for the creation of this committee was the suggestion by the forensic community that since 1992 there had been new scientific studies and data collected which obviated the need for many of the first NRC's recommendations. An important part of this new research were a number of studies examining statistical independence itself. Many of these studies have been previously cited in this paper. Since these studies tend to come to very different conclusions an important role for the new NRC committee would have been to review this research and provide its own scientific evaluations of the pros and cons of the competing claims.

In fact the new committee does consider these studies (NRC, 1996, chapter 4). However, the analysis is superficial and incomplete. While they clearly are disposed to accept the studies which have supported claims of independence they don't offer any scientific evaluation for why they support one set of hypotheses tests and not another. The most striking example of published papers which reach opposite conclusions regarding independence are Weir (1992) and Slimowitz and Cohen (1993) since they both analyzed exactly the same databases but come to different conclusions. The NRC report not only fails to analyze the different methodologies and conclusions in these papers but the report even fails to cite Slimowitz and Cohen as a paper making an important contribution to this issue.

When assessing the role of population substructure the committee uncritically accepts the inferences from the FBI's world wide study The problems with this type of analysis have already been discussed. While the committee does acknowledge that subgroups exist and may cause results from the product rule to be unreliable the committee offers only a partial solution. They suggest the incorporation of Wright's F-statistic when computing frequencies of homozygotes from PCR tests. This recommendation can correct for departures from Hardy-Weinberg equilibrium but not linkage equilibrium. The committee assumes departures from linkage equilibrium will be negligible.

The committee also correctly discusses a glaring and important design flaw in the FBI's fixed bin system (NRC, 1996, chapter 5; Fung, 1996). They point out that the size of many of the FBI's fixed bins are substantially less than their match criteria and thus biased to produce statistics that suggest matching profiles are rarer than they really are. To remedy this problem the first NRC committee had made the very reasonable suggestion that in those cases were calculations depend on fixed bins which are too small that they be made larger by merging them to a neighboring bin. The second NRC committee concludes this solution is not needed since, hopefully, profiles will consist of bins which are also larger than they need to be thus canceling out any errors created by bins which are too small. Since there is no guarantee that a DNA profile will always contain the appropriate mixture of bins which are too small and too large its hard to justify abandoning the first NRC's recommendation.

The committee also does not feel it is appropriate to use proficiency tests to attempt to estimate laboratory error rates (pgs. 3-10- 3-11). It appears that this conclusion is based on two different sorts of logic: (1) no proficiency test will ever fully replicate all the unusual features of real case work and hence are not relevant to real case work and (2) although several false positives were made in 1988 and 1989 in proficiency tests, since then no documented false positives have occurred, this signaling a time dependent trend in which error rates have now gone to zero or very close to it and thus can be safely ignored.

There are many compelling reason to reject these arguments and accept the original NRC's call to use proficiency tests to estimate laboratory error rates, here I review only some of the most obvious. If the NRC's logic about testing were correct than it should be impossible to assess the general level of competence of a lawyer by his performance on a bar exam since these involve the discussions of cases and facts which will certainly always differ from the details of "real' cases he will later deal with. In fact since most forensic proficiency tests are much less challenging than case work, error rates from proficiency tests could be considered a lower bound on case work error rates.

The arguments about the decline in error rates over time is also difficult to take seriously. As discussed earlier only if laboratory error rates were less than one in millions do they become unimportant. It seems unlikely that error rates, which may have been as high as 1 in 50 at Cellmark in late 1980's, have now dropped to 1 in millions. There is certainly no objective evidence to support this claim. Very recently a false match has been identified in actual case work. In late 1995 Cellmark mistakenly identified a know sample from a victim as the known sample from a defendant. This error than lead to the mistaken conclusion that the suspect matched DNA in an evidence sample which in fact only matched the victim (testimony of C. Word at California v. John Ivan Kocak, 17 November 1995, No. SCD110465).

A second example is found in testimony of Ms. Donna Dowden, a forensic scientist with the California Department of Justice DNA laboratory (California v. Noah Isaiah Wright, SC-078796A). In this testimony Ms. Dowden describes a false positive she made in an internal proficiency test utilizing DQ-$\alpha$. In addition to Ms. Dowden's error in her report was an error by the personnel who reviewed the report since they failed to detect the error the first time the report was reviewed. In this proficiency test the false match was actually due to an error in transcribing data from the original tests results to the final report (and of course of the supervisors failure to detect the error). Nevertheless, when Ms. Dowden was asked in court about her performance on proficiency tests she stated that she had performed satisfactorily. The basis for this opinion is Ms. Dowden's belief that since the molecular biological part of the tests gave the correct result her inability to correctly report this result was of no consequence. In fact Cellmark employee's have also suggested that the false match in the Koncak case does not constitute a false positive since the source of the error was clerical rather than a failure in the molecular biology (testimony of C. Word, California v. Bishop, 21 August 1996). Ultimately, a laboratory report which incorrectly declares a match between evidence and defendant DNA has the same misleading effect, whether the error is clerical in nature or represents a failure of the basic scientific procedures.

Even if one thinks proficiency tests can only provide rough estimates of laboratory error rates, a range of error rates could be utilized to show how they effect the ultimate weight of a DNA match. The failure of the second NRC report to make reasonable recommendations about estimating and reporting laboratory error rates has now been criticized repeatedly (Balding, 1997; Koehler, 1997; Lempert, 1997; Thompson, 1997).

There continues to exist a diversity of opinions on the proper manner to summarize the weight of DNA evidence. In attempting to respond to some of the criticisms of the first NRC report the second committee has contradicted several very useful recommendations in the first report. The application of DNA typing techniques in forensic science requires a balance between reasonable inference and tenuous speculation. That balance has not been achieved in the second report of the National Research Council.

**Acknowledgements.** I thank an anonymous referee for helpful comments on the manuscript and Dr. Paul Lewis for use of his software to carry out the exact tests.

REFERENCES

[1] Balding, D.J., *Errors and misunderstandings in the second NRC report*, Jurimetrics 37:469–476, 1997.

[2] Balzas, I., J. Neuweiler, P. Gunn, J. Kidd, K.K. Kidd, J. Kuhl, and L. Mingjun, *Human population genetic studies using hypervariable loci 1. Analysis of Assamese, Australian, Cambodian, Caucasian, Chinese and Melanesian populations*, Genetics 131:191–198, 1992.

[3] Brown, A.H.D. and M.W. Feldman, *Population structure of multilocus associations*, Proc. Natl. Acad. Sci. USA 78: 5913–5916, 1981.

[4] Budowle, B., A.M. Guisti, J.S. Waye, F.S. Baechtel, R.M. Fourney, D.E. Adams, L.A. Presley, H.A. Deadman, and K.L. Monson, *Fixed bin analysis for statistical evaluation of continuous distribution of allelic data from VNTR loci, for use in forensic comparisons*, Amer. J. Hum. Genet. 48:841–855, 1991.

[5] BUDOWLE, B., J.A. LINDSEY, J.A. DECOU, B.W. KOONS, S.M. GUISTI, C.T. COMEY, *Validation and population studies of the loci LDLR, GYPA, HBGG, D7S8 and Gc (PM loci), and HLA-DQα using a multiplex amplification and typing procedure*, J. Forensic Sci. 40:45–54, 1995.

[6] BUDOWLE, B., K.L. MONSON, A.M. GIUSTI, AND B.L. BROWN, *The assessment of frequency estimates of Hae III-generated VNTR profiles in various reference databases*, J. Forensic Sci. 39:319–352, 1994a.

[7] BUDOWLE, B., K.L. MONSON, A.M. GIUSTI, AND B.L. BROWN, *Evaluation of Hinf I generated VNTR profile frequencies determined using various ethni databases*, J. Forensic Sci. 39:988–1008, 1994b.

[8] CHAKRABORTY, R. AND S.P. DAIGER, *Polymorphisms at VNTR loci suggest homogeneity of the white population of Utah*, Human Biology 63:571–587, 1991.

[9] COHEN. J., M. LYNCH AND C.E. TAYLOR, *Forensic DNA tests and Hardy-Weinberg equilibrium*, Science 253:1037–1038, 1991.

[10] DEKA, R., R. CHAKRABORTY, AND R.E. FERRELL, *A population genetic study of six VNTR loci in three ethnically defined populations*, Genomics 11:83–92, 1991.

[11] DEVLIN, B. AND N. RISCH, *A note on Hardy-Weinberg equilibrium of VNTR data by using the Federal Bureau of Investigation's fixed-bin method*, Am. J. Hum. Genet. 51:549–553, 1992.

[12] DEVLIN, B., N. RISCH, AND K. ROEDER, 1990. *No excess of homozygosity at loci used for DNA fingerprinting*, Science 249:1416–, 1990.

[13] DEVLIN, B., N. RISCH, AND K. ROEDER, *Forensic DNA tests and Hardy-Weinberg equilibrium*, Science 253:1039–1041, 1991.

[14] EDWARDS, A.H.A. HAMMOND, L. JIN, T. CASKEY, AND R. CHAKRABORTY, *Genetic variation at five trimeric and tetrameric tandem repeat loci in four human population groups*, Genomics 12:241–253, 1992.

[15] FELDMAN, M.W., I. FRANKLIN, AND G.J. THOMSON, *Selection in complex genetic systems I. The symmetric equilibria of the three-locus symmetric viability model*, Genetics 76:135–162, 1974.

[16] FUNG, W.K., *A numeric 10% or 5% match window in DNA profiling ?* Forensic Sci. Int. 78:111–118, 1996.

[17] GEISSER, S., *Some statistical issues in forensic DNA profiling*, In: *Modelling and Prediction: honoring Seymour Geisser*, J.C. Lee, W.O. Johnson, A. Zellned (eds.), Springer, New York, 1996.

[18] GEISSER, S. AND W. JOHNSON, *Testing independence of fragment lengths within VNTR loci*, Am. J. Hum. Genet. 53:1103–1106, 1993.

[19] GEISSER, S. AND W. JOHNSON, *Testing independence when the form of the bivariate distribution is unspecified*, Statistics in Medicine 14:1621–1639, 1995.

[20] GREEN, P. AND E.S. LANDER, *Forensic DNA tests and Hardy-Weinberg equilibrium*, Science 253:1038–1039, 1991.

[21] GUO, S.W. AND E.A. THOMPSON, *Performing the exact test of Hardy-Weinberg proportion for multiple alleles*, Biometrics 48:361–372, 1992.

[22] HAGERMAN, P.J., *Letter to the editor*, Am. J. Hum. Genet. 47:876–877, 1990.

[23] HARTL, D.L. AND A.G. CLARK, *Principles of Population Genetics*, 2nd Ed., Sunderland, MA., Sinauer Associates, 1989.

[24] KIDD, J.R., F.L. BLACK, K.M. WEISS, I. BALAZS, AND K.K. KIDD, *Studies of three Amerindian populations using nuclear DNA polymorphisms*, Human Biology 63:775–794, 1991.

[25] KOEHLER, J.J., *Error and exaggeration in the presentation of DNA evidence at trial*, Jurimetrics 34:21–39, 1993.

[26] KOEHLER, J.J., *Why DNA likelihood ratios should account for error (even when a National Research Council Report say they should not)*, Jurimetrics 37:425–437, 1997.

[27] KOEHLER, J.J., A. CHIA AND S. LINDSEY, *The random match probability (RMP) in DNA evidence: irrelevant and prejudicial ?*, Jurimetrics, 35:201–219, 1995.

[28] KRANE, D.E., R.W. ALLEN, S.A. SAWYER, D.A. PETROV AND D.L. HARTL, *Genetic differences at four DNA typing loci in Finnish, Italian and mixed Caucasian populations*, Proc. Natl. Acad. Sci. USA 89:10583–10587, 1992.

[29] LANDER, E. AND B. BUDOWLE, *DNA fingerprinting dispute laid to rest*, Nature 371:735–738, 1994.

[30] LEMPERT, R., *Some caveats concerning DNA as criminal identification evidence: with thanks to the Reverend Bayes*, Cardozo Law Review 13:303–341, 1991.

[31] LEMPERT, R., *After the DNA wars : skirmishing with NRC II*, Jurimetrics 37:439–468, 1997.

[32] LEWONTIN, R.C. AND D.L. HARTL, *Population genetics in forensic DNA typing*, Science 254:1745–1750, 1991.

[33] MAISTE, P.J. AND B.S. WEIR, *A comparison of tests for independence in the FBI RFLP data bases*, Genetica 96:125–138, 1995.

[34] MILLER, R.G., *Simultaneous Statistical Inference*, McGraw-Hill: New York, 1966.

[35] MORTON, N.E., *Genetic structure of forensic populations*, Proc. Natl. Acad. Sci. USA 89:2556–2560, 1992.

[36] MUELLER, L.D., *Population genetics of hypervariable human DNA*, In: *Forensic DNA Technology*, M.A. Farley, J.J. Harrington (eds.), Lewis Publishers, Chelsea, MI, 1991.

[37] MUELLER, L.D., *The use of DNA typing in forensic science*, Accountability in Research 3:55–67, 1993.

[38] MUELLER, L.D., *Letter to the Editor*, J. Forensic Sci. 43:446–447, 1998.

[39] National Research Council, *DNA Technology in Forensic Science* , National Academy Press, Washington, D. C., 1992.

[40] National Research Council *The Evaluation of Forensic DNA Evidence*, National Academy Press, Washington, D. C., 1996.

[41] ODELBERG, S.J., R. PLATKE, J.R. ELDRIDGE, L. BALLARD, P. O'CONNELL, Y. NAKAMURA, M. LEPPERT, J.M. LALOUEL, AND R. WHITE, *Characterization of eight VNTR loci by agarose gel electrophoresis*, Genomics 5:915–924, 1989.

[42] RISCH, N.J. AND B. DEVLIN, *On the probability of matching DNA fingerprints*, Science 255:717–720, 1992.

[43] SAJANTILA, A., B. BUDOWLE, M. STROM, V. JOHNSSON, M. LUKKA, L. PELTONEN, AND C. EHNHOLM, *PCR amplification of alleles at the D1S80 locus: comparison of a Finnish and a North American Caucasian population sample, and forensic casework evaluation*, Am. J. Hum. Genet. 50:816–825, 1992.

[44] SAWYER, S., D. KRANE, A. PODLESKI AND D. HARTL, *DNA fingerprinting loci do show population differences - comments on Budowle et al.*, Am. J Hum. 59:272–274, 1996.

[45] SLIMOWITZ, J.R. AND J.E. COHEN, *Violations of the ceiling principle: exact conditions and statistical evidence*, Am. J. Hum. Genet. 53:314–323, 1993.

[46] THOMPSON, W.C., *Subjective interpretation, laboratory error, and the value of forensic DNA evidence: three case studies*, Genetica 92:153–168, 1995.

[47] THOMPSON, W.C., *Accepting lower standards: the National Research Council's second report on forensic DNA evidence*, Jurimetrics 37:405–424, 1997.

[48] WEIR, B.S., *Independence of VNTR alleles defined as fixed bins*, Genetics 130:873–887, 1992a.

[49] WEIR, B.S., *Independence of VNTR alleles defined as floating bins*, Am. J. Hum. Genet. 51:992–997, 1992b.

[50] ZAYKIN, D., L. ZHIVOTOVSKY AND B.S. WEIR, *Exact tests for association between alleles at arbitrary numbers of loci*, Genetica 96:169–178, 1995.

# IN DISPRAISE OF INCONSISTENT DNA RELATIVE FREQUENCY ESTIMATES

SEYMOUR GEISSER*

**1. Introduction.** In a series of papers, Geisser and Johnson (1992, 1993) and Geisser (1996) criticized some of the statistical methods employed and the databases gathered by the Federal Bureau of Investigation (FBI) and Cellmark, two of the principal forensic laboratories engaged in Restricted Fragment Length Polymorphisms (RFLP) DNA matching and profiling. The major criticisms were: (a) lack of random sampling and of adequate sample sizes, e.g. the fact that Cellmark's small Black database was obtained solely from rare blood donor volunteers at a Detroit blood bank and similarly their Caucasian database from the Blood Bank of Delaware which was used throughout the United States; (b) lack of mutual independence among alleles within loci and between loci; (c) the fact that the product rule(the multiplication of probabilities of independent events), even under the appropriate assumptions, i.e. random sampling and mutual independence, will underestimate the relative frequency of a profile; (d) the lack of blind, external and periodic proficiency tests; and (e) the fact that the FBI 5% window for the original match led to a theoretically computable false exclusion rate for 4 probes of between $10^{-8}$ and $10^{-10}$ without any notion of what the false inclusion rate would be. In addition the FBI claimed that their use of fixed bins, to estimate the relative frequency of a DNA profile, to be conservative in that it would overestimate the profile relative frequency, Budowle (1991). Failure on the part of the National Research Council (NRC) report (1996) to criticize forensic laboratories for these flaws is inexcusable.

In what follows we take up another flawed feature of the FBI procedure and an inadequate feature of the databases that Cellmark has been using in cases throughout the United States.

**2. Inconsistency of initial match and database match.** An RFLP allelic band, though discreet, is measured in base pairs with error rendering actual measurements quasi-continuous. Hence a window of some sort in which to assign similar allelic band values is required in order to be able to decide on whether band values "match." An individual's profile consisting of several loci with two bands per locus is used to estimate the relative frequency of the profile in a population.

The FBI's initial window, to determine whether two bands match, is essentially to indicate a match if the band of the alleged is within ±5% of the crime scene band value. When this is the case the next step is to obtain an estimate of the relative frequency of the band in which a racial sample of

*School of Statistics, University of Minnesota, Minneapolis, MN 55455-0436.

individuals is distributed. Now instead of using a floating window as in the initial match the FBI uses a fixed binning procedure which initially had 31 bins. For each locus, estimates of the relative frequency of the two bands are determined by the fraction of the racial sample in the bin in which the alleged's band lies. These two fractions are multiplied together which is then doubled to obtain the estimated relative frequency for the locus. Then the product of the relative frequencies for the various loci probed is reported as the relative frequency of the individual profile. This "product rule" assumes independence within each locus and between loci.

The width of the FBI bins was criticized by Fung (1996) who demonstrated that more than half of these bins were less than ±5% of their midpoint in size, or of width 10%, belying the claim by Budowle of conservatism i.e. favoring the defendant. However, because of the distribution of bands and relatively small size of the database, the FBI has some time ago, for use in the United States, adjusted their binning procedure for forensic cases to between 13 and 24 bins depending on the locus being probed. This presumptively would favor the defendant since the bins are now wider on average and hence have higher relative frequencies.

A second critical issue, not addressed by Fung, is that the initial match is based on an intragel comparison (when the matched bands are on the same electrophoretic gel) while the match between the alleged or suspect profile and those in the database is necessarily an intergel (different gels) comparison. For bands less than 10,000 basepairs, the intergel standard deviation is approximately 1.82 times the intragel standard deviation. This was ascertained from data obtained in a court ordered discovery[1] of duplicate measurements on individuals made on the same gel and on different gels. This would indicate that, for the sake of consistency, a match of a suspect band value should include all database bands within a ±9.1% of the suspect band value.

We now investigate the size % of the bins for the Caucasian database, the largest of the several FBI databases, in terms of the number of bins above ±9.1% of midpoint band size and the fraction of the database residing in bins above this 18.2% width, see Table 1. From there, it is clear that very few bins are greater than this standard in width and the percentage of individuals in those bins is quite small, both in magnitude and in relation to proportion of bins below 18.2%.

Even if all other assumptions such as random sampling, mutual independence of alleles within and between loci are considered to obtain, permitting the use of the product rule, and a correction made for underestimation by the product rule, Geisser (1996), the FBI's claim to conservatism based on a consistent relationship between their initial match window and their binning procedure, is clearly false.

The differences between the percentages in the last two columns of Table 1 which reflect mainly end effects, i.e. the first and last bins for

[1] Minnesota vs. Johnson, Fourth Judicial District Court, 1991

each locus (essentially the tails of the distribution) are very large relative to the number in the bin. The last bin of course is infinite in size and hence always larger than 18.2%. If these two were eliminated, noting that the fraction of the individuals in the database is very small in these bins, the situation would belie even further the claim of conservatism, which was based on criteria inconsistently applied. If anything, it demonstrates that the FBI's estimates could appreciably underestimate the relative frequency of profiles without even considering several other flawed assumptions.

TABLE 1
*FBI Caucasian Database (Rebinned).*

| | (1) | (2) | (3) | (4) |
|---|---|---|---|---|
| Locus | No. of bins | No. > ±9.1% | % | % of database in (2) |
| D2S44 | 21 | 3 | 14.3 | 2.1 |
| D17S79 | 13 | 3 | 23.1 | 2.0 |
| D1S7 | 26 | 2 | 7.7 | 5.5 |
| D4S139 | 17 | 2 | 11.8 | 10.6 |
| D10S28 | 23 | 2 | 8.7 | 2.1 |
| D14S13 | 24 | 5 | 20.8 | 2.6 |

**3. Pairwise comparisons.** To have an idea of how much underestimation there might be, we can calculate in a database the number of matches involving all possible pairs of individual profiles. We shall use the FBI Caucasian database which is the largest of their racial databases to provide us with some idea of the extent of the underestimation. The five most commonly used RFLP loci will be investigated. A sixth D14S13 is less frequently used in court cases.

Two different estimates are compared in Table 2, (1) the pairwise direct count using a ±9.1% floating bin which does not rely on independence and (2) the FBI adjusted binning procedure giving the 5, 25, 50, 75 and 95 percentile values using the product rule as the FBI does. The relative frequency for the direct count varies in the range of 1 in 4,700–53,000 for 3 loci matches. For 4 loci it varies in the range of 1 in 70,000–80,000. There was also one 5-probe match out of the 73,000 odd comparisons and that match turned out to be a six-probe match if D14S13 were included! These values are quite different from what the use of the FBI bins and procedure yield and certainly different, on average, from the most frequently reported values of 1 in many millions and billions in court cases. The estimates in Table 2 are clearly underestimates since matches using the narrower FBI criterion were deleted by the FBI from the original data set. Some were deleted because the individual was later discovered to be profiled twice but others were deleted for no reason other than the application of the FBI criterion itself.

TABLE 2

*Relative frequency in FBI Caucasian database for 3 and 4 loci using a ±9.1% window, pairwise direct count, and FBI bins using product rule.*

| 3 Locus Matches (Omitted Loci) | No. of Individuals | Relative frequency (1 out of): Pairwise Direct Count | Product Rule: Percentiles using FBI Bins | | | | |
|---|---|---|---|---|---|---|---|
| | | ±9.1% | 95% | 75% | 50% | 25% | 5% |
| D10S28, D17S79 | 565 | $5.3 \times 10^4$ | $1.8 \times 10^5$ | $5.9 \times 10^5$ | $1.7 \times 10^6$ | $5.1 \times 10^6$ | $2.6 \times 10^7$ |
| D17S79, D4S139 | 410 | $4.2 \times 10^4$ | $6.3 \times 10^5$ | $1.7 \times 10^6$ | $4.9 \times 10^6$ | $1.2 \times 10^7$ | $5.2 \times 10^7$ |
| D4S139, D10S28 | 548 | $1.0 \times 10^4$ | $5.3 \times 10^4$ | $1.3 \times 10^5$ | $3.9 \times 10^5$ | $1.4 \times 10^6$ | $1.0 \times 10^7$ |
| D1S7, D17S79 | 405 | $4.1 \times 10^4$ | $1.8 \times 10^5$ | $5.5 \times 10^5$ | $1.7 \times 10^6$ | $4.9 \times 10^6$ | $2.5 \times 10^7$ |
| D10S28, D1S7 | 549 | $4.7 \times 10^3$ | $1.5 \times 10^4$ | $4.6 \times 10^4$ | $1.2 \times 10^5$ | $4.5 \times 10^5$ | $5.8 \times 10^6$ |
| D4S139, D1S7 | 401 | $5.7 \times 10^3$ | $5.4 \times 10^4$ | $1.2 \times 10^5$ | $3.0 \times 10^5$ | $1.2 \times 10^6$ | $9.7 \times 10^6$ |
| D2S44, D17S79 | 402 | $4.0 \times 10^4$ | $3.2 \times 10^5$ | $9.8 \times 10^5$ | $2.9 \times 10^6$ | $8.3 \times 10^6$ | $3.8 \times 10^7$ |
| D2S44, D10S28 | 543 | $1.1 \times 10^4$ | $2.7 \times 10^4$ | $7.4 \times 10^4$ | $2.0 \times 10^5$ | $9.2 \times 10^5$ | $9.0 \times 10^6$ |
| D2S44, D4S139 | 396 | $1.3 \times 10^4$ | $9.2 \times 10^4$ | $2.1 \times 10^5$ | $5.4 \times 10^5$ | $2.0 \times 10^6$ | $1.6 \times 10^7$ |
| D2S44, D1S7 | 397 | $7.9 \times 10^3$ | $2.6 \times 10^4$ | $7.0 \times 10^4$ | $1.8 \times 10^5$ | $8.9 \times 10^5$ | $6.1 \times 10^6$ |
| **4 Locus Matches** (Locus Omitted) | | | | | | | |
| D17S79 | 396 | $7.8 \times 10^4$ | $2.4 \times 10^7$ | $1.1 \times 10^8$ | $3.6 \times 10^8$ | $1.1 \times 10^9$ | $6.3 \times 10^9$ |
| D10S28 | 535 | $7.1 \times 10^4$ | $2.1 \times 10^6$ | $8.8 \times 10^6$ | $2.7 \times 10^7$ | $1.1 \times 10^8$ | $1.6 \times 10^9$ |
| D4S139 | 391 | $7.6 \times 10^4$ | $6.7 \times 10^6$ | $2.4 \times 10^7$ | $7.5 \times 10^7$ | $2.9 \times 10^8$ | $2.1 \times 10^9$ |
| D1S7 | 391 | $7.6 \times 10^4$ | $2.0 \times 10^6$ | $7.5 \times 10^6$ | $2.5 \times 10^7$ | $9.7 \times 10^7$ | $1.3 \times 10^9$ |
| D2S44 | 387 | $7.5 \times 10^4$ | $3.7 \times 10^6$ | $1.3 \times 10^7$ | $4.0 \times 10^7$ | $2.2 \times 10^8$ | $1.9 \times 10^7$ |

The second NRC report (1996) also quotes a similar study done by Risch and Devlin (1992) on pairs using the direct count. They reported no matches in 4 or 5 probes among 7.6 million pairs of profiles. But Risch and Devlin used a ±2.4% window so these results were quite inconsistent with even the FBI initial ±5% window. The NRC report also stated, based on a personal communication from R. Chakraborty which remains unpublished, that out of 58 million pairwise comparisons, only 2 matches were found. No mention is made of the match criterion used.

While we agree with the NRC report that the same criterion (a floating bin) should be used for the initial match and the database match we assert that it should be adjusted for the difference in variation inherent in an intragel and intergel comparison. It is clear from Table 2 that the inconsistent FBI procedure from a 4 locus match could underestimate the relative frequency in the range of $10^3$–$10^5$ or more. Calling such a procedure conservative or as favoring the defendant defies common sense, since it is assumed that the smaller the relative frequency the stronger the evidence against the accused.

With regard to databases that were used by Cellmark, not only are the totals inadequate, see Table 3, but the number of individuals that have measurements on all five of their probes are 2 for Blacks, 59 for Hispanics, and 75 for Caucasians. This demonstrates the inadequacy for rigorous testing of linkage equilibrium or for a comparable direct count pairwise assessment of the individual profiles. For the inadequacy of the databases containing individuals measured on 4 loci out of 5 loci, see Table 4.

TABLE 3
*Number of individuals with values on Cellmark Loci.*

| Locus | Blacks | Caucasians | Hispanics |
|---|---|---|---|
| D1S7 | 240 | 262 | 215 |
| D7S21 | 238 | 264 | 183 |
| D12S11 | 223 | 294 | 192 |
| D7S22 | 200 | 325 | 168 |
| D2S44 | 146 | 208 | 110 |

TABLE 4
*Number of individuals with values on 4 Loci.*

| Omitted Locus | Blacks | Caucasians | Hispanics |
|---|---|---|---|
| D1S7 | 2 | 79 | 63 |
| D7S21 | 2 | 108 | 73 |
| D12S11 | 2 | 77 | 72 |
| D7S22 | 31 | 91 | 76 |
| D2s44 | 8 | 153 | 109 |

**Acknowledgment.** This work was supported in part of NIGMS grant 25271.

## REFERENCES

[1] BUDOWLE, B. ET AL., Fixed Bin Analysis for Statistical Evaluation of Continuous Distribution of Allelic Data from VNTR Loci for Use in Forensic Comparisons. *American Journal of Human Genetics* **48**, 841–855 (1991).

[2] FUNG, W.K., 10% or 5% match window in DNA profiling. *Forensic Sci. Int.* **78**, 111–118 (1996).

[3] GEISSER, S., Some statistical issues in forensic DNA. In *Modelling and Prediction*, J.C. Lee et al. (eds.), Springer-Verlag, New York, 3–18 (1996).

[4] GEISSER, S. AND JOHNSON, W., Testing Hardy-Weinberg equilibrium on allelic data from VNTR loci. *American Journal of Human Genetics* **53**, 1084–1089 (1996).

[5] GEISSER, S. AND JOHNSON, W., Testing independence of fragment lengths within VNTR loci. *American Journal of Human Genetics* **53**, 1103–1106 (1996).

[6] National Research Council, *National Research Council Report: The Evaluation of Forensic DNA Evidence, National Academy Press* (1996).

[7] RISCH, N.J. AND DEVLIN, B., On the probability of matching fingerprints. *Science* **255**, 717–720 (1992).

# QUANTIFYING THE GENETIC STRUCTURE OF POPULATIONS WITH APPLICATION TO PATERNITY CALCULATIONS

B.S. WEIR*

**1. Introduction.** The usual situation in paternity testing is that mother, child and alleged father are genotyped. The alleged father is declared "not excluded" if he carries an allele that is inferred to be the child's paternal allele. Beyond that, in many states there is a legal requirement for calculation of a "paternity index" or of a "probability of paternity." Both these require a likelihood ratio, based on the probabilities of the genetic evidence $E$ under alternative explanations:

$H_p$: The alleged father is the father.

$H_d$: The alleged father is not the father.

From Bayes' theorem

$$\frac{\Pr(H_p|E)}{\Pr(H_d|E)} = \frac{\Pr(E|H_p)}{\Pr(E|H_d)} \times \frac{\Pr(H_p)}{\Pr(H_d)}$$

where $\Pr(H_p)$ and $\Pr(H_d)$ are the prior probabilities, i.e., prior to the genetic evidence. The likelihood ratio is generally referred to as the paternity index, and will be written here as PI:

$$\mathrm{PI} = \frac{\Pr(E|H_p)}{\Pr(E|H_d)}$$

If the (posterior) probability of paternity is written as $\pi = \Pr(H_p|E)$, then

$$\frac{\pi}{1-\pi} = \mathrm{PI} \times \frac{\Pr(H_p)}{\Pr(H_d)}$$

If prior probability of paternity $\Pr(H_p)$ is set to 0.5, then

$$\pi = \frac{\mathrm{PI}}{1+\mathrm{PI}}$$

Although there is little scientific justification for such a prior, it is part of several state statutes.

The PI can be expressed in terms of the probability of genotype $G_C$ of the child **C**, conditional on the genotype $G_M$ of the mother **M** and on

*Program in Statistical Genetics, Department of Statistics, North Carolina State University, Raleigh, NC 27695-8203.

the genotype $G_A$ of the alleged father **A**:

$$\begin{aligned} \mathrm{PI} &= \frac{\Pr(E|H_p)}{\Pr(E|H_d)} \\ &= \frac{\Pr(G_C|G_M,G_A,H_p)}{\Pr(G_C|G_M,G_A,H_d)} \times \frac{\Pr(G_M,G_A|H_p)}{\Pr(G_M,G_A|H_d)} \\ &= \frac{\Pr(G_C|G_M,G_A,H_p)}{\Pr(G_C|G_M,G_A,H_d)} \end{aligned}$$

since the genotypes of mother and alleged father are independent of $H_p$ and $H_d$.

Suppose genotyping has been conducted for some locus **a** for which the alleles are written $a_i$. The numerator of PI follows from Mendelian rules, e.g., if the child is heterozygous $a_ia_j$, the mother homozygous $a_ia_i$, and the alleged father heterozygous $a_ja_k$:

$$\Pr(G_C = a_ia_j|G_M = a_ia_i, G_A = a_ja_k, H_p) = 0.5$$

Throughout this discussion, distinct alleles $a_i$ at a locus **a** will be indicated by distinct subscripts, so that $a_i$ and $a_j$ are distinguishable alleles - they are not alike in state.

It is convenient to work with the maternal and paternal alleles $A_M$ and $A_P$ that make up the child's genotype: $G_C = A_MA_P$. This allows the expression

$$\Pr(G_C|G_M,G_A,H) = \Pr(A_M|G_M,H)\Pr(A_P|A_M,G_M,G_A,H)$$

Because it is being assumed that the child's mother is known and has been genotyped, Mendelian laws provide

$$\begin{aligned} \Pr(A_M = a_i|G_M = a_ia_i, H) &= 1 \\ \Pr(A_M = a_i|G_M = a_ia_j, H) &= 0.5 \end{aligned}$$

and these hold for both $H_p$ and $H_d$.

Under explanation $H_p$, the alleged father has provided the paternal allele so

$$\begin{aligned} \Pr(A_P = a_i|A_M,G_M,G_A = a_ia_i,H_p) &= 1 \\ \Pr(A_P = a_i|A_M,G_M,G_A = a_ia_j,H_p) &= 0.5 \\ \Pr(A_P = a_i|A_M,G_M,G_A = a_ja_k,H_p) &= 0 \end{aligned}$$

Under explanation $H_d$, however, some other man **T** has provided the paternal allele and the conventional treatment sets

$$\Pr(A_P = a_i|A_M,G_M,G_A,H_d) = p_i$$

for all values of $G_A$, where $p_i$ is the proportion of alleles in the population that are of type $a_i$. Note that the other man is generally unknown, so his genotype $G_T$ is not known. The PI takes the values shown in Table 1.

TABLE 1
***Non-zero PI values when there is no relationship among mother, father and alleged father.***

| Paternal allele | Alleged father | Paternity index |
|---|---|---|
| $a_i$ | $a_i a_i$ | $\frac{1}{p_i}$ |
| | $a_i a_j$ | $\frac{1}{2p_i}$ |

**2. Effects of relatedness.** In general, the probability of $A_P$ under explanation $H_d$ depends on $G_M$ and $G_A$. In particular, the relationship among alleles brought about by a common evolutionary history for individuals in a population imposes a dependence. The degree of dependence can be expressed in terms of measures of identity by descent, ibd: alleles that descend from the same ancestral allele are said to be ibd. The probabilities with which two alleles are ibd can be defined for the two separate arrangements of the alleles: within or between individuals. For individual **A** who has alleles $a$ and $a'$, the descent measure is called the inbreeding coefficient and can be defined as

$$F_A = \Pr(a \equiv a'), \ a, a' \text{ in } A$$

with the equivalence sign indicating ibd. If $a$ is a random one of the two alleles of individual **A**, and $a'$ a random one of the two alleles of individual **T**, the descent measure for $a$ and $a'$ is called the coancestry of **A** and **T**:

$$\theta_{AT} = \Pr(a \equiv a'), \ a, a' \text{ random alleles from } A, T$$

This probability interpretation dates to Malécot (1969), and the notation used here is that of Cockerham (1971).

Originally, Wright (1921) defined inbreeding coefficients in terms of increase in homozygosity over that expected for random-mating populations. He expressed the population proportions of genotypes $a_i a_i$ and $a_i a_j$ in terms of the population $F$ and allele proportions:

$$a_i a_i : P_{ii} = F p_i + (1 - F) p_i^2$$
$$a_i a_j : P_{ij} = 2(1 - F) p_i p_j, \ i \neq j$$

where $F$ holds for a random member of the population. It is generally assumed that $F$ is the same for all members of a population. A parallel formulation holds for the probabilities of alleles $a_i, a_i$ or $a_i, a_j$ being carried by two random individuals. These probabilities are written as $P_{i,i}$ and $P_{i,j}$, and

$$\begin{aligned} a_i, a_i &: P_{i,i} = \theta p_i + (1-\theta)p_i^2 \\ a_i, a_j &: P_{i,j} = 2(1-\theta)p_i p_j,\ i \neq j \end{aligned}$$

Wright's formulations did not require $F$ and $\theta$ to be positive, but his equations can be derived for $F$ and $\theta$ defined as ibd probabilities.

Suppose maternal allele $A_M$ and paternal allele $A_P$ can be determined from genotypes $G_M$ and $G_C$. Paternity calculations require $\Pr(A_P|G_M, G_A, H_d)$, the probability of $A_P$ being a random allele in man **T** conditional on the genotypes of the mother **M** and the alleged father **A**. From the definition of conditional probabilities

$$\Pr(A_P|G_M, G_A, H_d) = \frac{\Pr(A_P, G_M, G_A|H_d)}{\Pr(G_M, G_A|H_d)}$$

showing that it is necessary to consider the relationships among sets of five alleles: the paternal allele $A_P$ and the two alleles in each of the mother and the alleged father. If the population can be assumed to have reached an evolutionary equilibrium, then allele proportions satisfy a Dirichlet distribution (Wright, 1951), and simple expressions are available for the joint probabilities of sets of alleles. This treatment is purely allelic, meaning that no account is taken of the arrangement of alleles within and among individuals. In particular, $F = \theta$ within such a population, and it is usual to express results in terms of $\theta$. The probability for a set of alleles, in which allele $a_i$ occurs $t_i$ times, is

$$\Pr(\prod_i a_i^{t_i}) = \frac{\Gamma(g.)}{\Gamma(t. + g.)} \prod_i \frac{\Gamma(t_i + g_i)}{\Gamma(g_i)}$$

where

$$\begin{aligned} g_i &= (1-\theta)p_i/\theta \\ g. &= \sum_i g_i = (1-\theta)/\theta \\ t. &= \sum_i t_i \end{aligned}$$

Wright (1951) found the Dirichlet held for stationary populations, whereas Crow and Kimura (1970) and Griffiths (1979) treated non-stationary populations. Further discussion has been given by Jiang (1987) and Li (1996). A complete set of PI values for stationary populations to which $\theta$ applies is shown in Table 2. They have been given previously by Balding and Nichols (1995).

**3. Meaning of $\theta$.** The parameter $\theta$ used in Table 2 is for a random pair of alleles within one population, and it refers to the relationship of those alleles relative to the relationship between alleles in different populations. This concept of relativity has implications for methods used to estimate $\theta$, and it will now be amplified. Although the methodology is intended for human populations, the meaning of "relative to" is most easily explained for the simplest possible mating system of selfing. An individual that has arisen by selfing has received both maternal and paternal alleles from a single parent, and many plant species are capable of both selfing and outcrossing.

**3.1. Selfing example.** Suppose a large population consists of unrelated and non-inbred individuals. Suppose further that the proportions of alleles $a_1$ and $a_2$ at locus **a** are both 0.5, so that the population proportions of the three genotypes $a_1a_1, a_1a_2$, and $a_2a_2$ are 0.25, 0.50, and 0.25. If each individual selfs and produces a large number of offspring, these offspring sets provide three types of subpopulations within the offspring population:

| Type | Proportion | Composition |
|---|---|---|
| 1 | $\frac{1}{4}$ | $a_1a_1$ |
| 2 | $\frac{1}{2}$ | $\frac{1}{4}a_1a_1 + \frac{1}{2}a_1a_2 + \frac{1}{4}a_2a_2$ |
| 3 | $\frac{1}{4}$ | $a_2a_2$ |

Within each subpopulation, the genotypic proportions obey the Hardy-Weinberg rule, meaning that genotype proportions are products of allele proportions:

$$\begin{aligned}
a_1a_1 &: P_{11} = p_1^2 \\
a_1a_2 &: P_{12} = 2p_1p_2 \\
a_2a_2 &: P_{22} = p_2^2
\end{aligned}$$

Frequencies $(p_1, p_2)$ take the values (1,0), (0.5,0.5), and (0,1) in the three types of subpopulation. Moreover, within each subpopulation each pair of alleles is equally related - whether they are within the same or different offspring.

A more general formulation parameterizes genotype proportions within each subpopulation in terms of the within-subpopulation inbreeding coefficient $f$:

$$\begin{aligned}
a_1a_1 &: P_{11} = p_1^2 + fp_1(1-p_1) \\
a_1a_2 &: P_{12} = 2p_1p_2(1-f) \\
a_2a_2 &: P_{22} = p_2^2 + fp_2(1-p_2)
\end{aligned}$$

Proportions $p_1$ and $p_2$ are within a subpopulation and $f = 0$ in this example.

However, if the subpopulation boundaries are ignored, the allele proportions in the whole offspring population are $p_1 = p_2 = 0.5$ as they were among the parents, and the genotype proportions for $a_1a_1, a_1a_2, a_2a_2$ depart from the parental 1:2:1 ratio. Population-wide genotype proportions are parameterized with the total inbreeding coefficient $F$:

$$\begin{aligned} a_1a_1 &: P_{11} = p_1^2 + Fp_1(1-p_1) \\ a_1a_2 &: P_{12} = 2p_1p_2(1-F) \\ a_2a_2 &: P_{22} = p_2^2 + Fp_2(1-p_2) \end{aligned}$$

For alleles carried by different offspring in the same subpopulation, the proportions *averaged over subpopulations* are

$$\begin{aligned} a_1, a_1 &: P_{1,1} = p_1^2 + \theta p_1(1-p_1) \\ a_1, a_2 &: P_{1,2} = 2p_1p_2(1-\theta) \\ a_2, a_2 &: P_{2,2} = p_2^2 + \theta p_2(1-p_2) \end{aligned}$$

Frequencies $p_1 = 0.5$ and $p_2 = 0.5$ are averages over all subpopulations, and $F = \theta = 0.5$ because there is a 50% probability that any two offspring alleles within the same subpopulation derive from the same allele in the single parent for that subpopulation. The overall $a_1a_1$ proportion in the population is 3/8, although this is not the proportion in any one subpopulation.

A situation often faced in human applications is that the population is structured, reflecting some local restrictions on mating (although not as extreme as the selfing restriction of this example), and it is not known to which subpopulation any person belongs. The structuring is hidden, and the only information available is for the whole population. If the probability of an $a_1a_1$ genotype in one subpopulation is needed in such a situation, it cannot be found from $p_1^2$ when $p_1$ is estimated from the whole population. The appropriate $p_1$ for the subpopulation is not known, and the average $P_{11} = p_1^2 + Fp_1(1-p_1)$ is taken to apply to any subpopulation

A sample from the whole population provides an estimate of $p_1$. This is the mean over subpopulations. In the selfing example, if $p_{1s}$ is the proportion in the $s$th subpopulation, taking expectations over subpopulations gives

$$\begin{aligned} \mathcal{E}(p_{1s}) &= p_1 \\ \text{i.e. } \mathcal{E}(p_{1s}) &= \frac{1}{4} \times 1 + \frac{1}{2} \times \frac{1}{2} + \frac{1}{4} \times 0 \\ &= \frac{1}{2} \end{aligned}$$

The variance of allele proportions over subpopulations is

$$\begin{aligned}\mathrm{Var}(p_{1s}) &= \frac{1}{4}(1-\frac{1}{2})^2 + \frac{1}{2}(\frac{1}{2}-\frac{1}{2})^2 + \frac{1}{4}(0-\frac{1}{2})^2 \\ &= 1/8 \\ \text{i.e., } \mathrm{Var}(p_{1s}) &= \theta p_1(1-p_1)\end{aligned}$$

The degree of relationship of alleles within subpopulations, relative to the whole population, is part of the component of variance of allele proportions over subpopulations.

This last expression suggests a means for estimating $\theta$. A large-sample approximation uses the sample mean and variance of allele proportions:

$$\hat{\theta} = \frac{s_1^2}{\bar{p}_1(1-\bar{p}_1)}$$

The role of $\theta$ in quantifying the variance of allele frequencies over subpopulations also suggests that it can be used as a measure of distance, as shown in the next section.

For paternity calculations, Table 2 is interpreted as providing PI values when the mother, alleged father and true father all belong to the same subpopulation. This is the most conservative assumption in that it leads to smaller PI values than if population structuring is ignored or if the three people belong to different subpopulations. Allelic data are generally available only at the population level instead of at the level of a particular subpopulation, and $\theta$ describes allelic frequency variation among subpopulations. The effect of using Table 2 instead of the conventional Table 1 can be substantial, and can easily lead to a halving of the PI for realistic values of $\theta$, such as 0.03.

**4. $\theta$ as a distance.** The parameter $\theta$ has a key role to play in the formulation of paternity indices, but also it is central to evolutionary theory. When $\theta$ is defined as the ibd probability of any two alleles in a random mating population of size $N$, and there are no disturbing forces, it changes over time according to

$$\theta(t+1) = \frac{1}{2N} + \left(1-\frac{1}{2N}\right)\theta(t)$$

With an initial condition of $\theta(0) = 0$, therefore,

$$\begin{aligned}\theta(t) &= 1 - \left(1-\frac{1}{2N}\right)^t \\ &\approx t/2N \qquad (1)\end{aligned}$$

Although $\theta$ is defined for pairs of alleles within populations, it serves as a measure of distance between populations (Reynolds et al., 1983). This

TABLE 2

*Non-zero PI values when mother, father and alleged father all belong to the same subpopulation in which $\theta$ is the probability that two random alleles are identical by descent.*

| $G_C$ | $G_M$ | $G_A$ | $PI$ |
|---|---|---|---|
| $a_ia_i$ | $a_ia_i$ | $a_ia_i$ | $\frac{1+3\theta}{4\theta+(1-\theta)p_i}$ |
| | | $a_ia_j$ | $\frac{1+3\theta}{2[3\theta+(1-\theta)p_i]}$ |
| | $a_ia_j$ | $a_ia_i$ | $\frac{1+3\theta}{3\theta+(1-\theta)p_i}$ |
| | | $a_ia_j$ | $\frac{1+3\theta}{2[2\theta+(1-\theta)p_i]}$ |
| | | $a_ia_k$ | $\frac{1+3\theta}{2[2\theta+(1-\theta)p_i]}$ |

| $G_C$ | $G_M$ | $G_A$ | $PI$ |
|---|---|---|---|
| $a_ia_j$ | $a_ia_i$ | $a_ja_j$ | $\frac{1+3\theta}{2\theta+(1-\theta)p_j}$ |
| | | $a_ia_j$ | $\frac{1+3\theta}{2[\theta+(1-\theta)p_j]}$ |
| | | $a_ja_k$ | $\frac{1+3\theta}{2[\theta+(1-\theta)p_j]}$ |
| | $a_ia_j$ | $a_ia_i$ | $\frac{1+3\theta}{4\theta+(1-\theta)(p_i+p_j)}$ |
| | | $a_ia_j$ | $\frac{1+3\theta}{4\theta+(1-\theta)(p_i+p_j)}$ |
| | | $a_ja_k$ | $\frac{1+3\theta}{2[3\theta+(1-\theta)(p_i+p_j)]}$ |
| | $a_ia_k$ | $a_ja_j$ | $\frac{1+3\theta}{2\theta+(1-\theta)p_j}$ |
| | | $a_ja_k$ | $\frac{1+3\theta}{2[\theta+(1-\theta)p_j]}$ |
| | | $a_ja_l$ | $\frac{1+3\theta}{2[\theta+(1-\theta)p_j]}$ |

Different subscripts indicate different allelic types.

was illustrated in the selfing example, where $\theta$ described allele frequency variance over subpopulations, and this variance would increase if selfing continued over time. If $\theta$ is estimated from a set of populations (see next section) that diverged from an ancestral population $t$ generations ago, then it furnishes an estimate of this time scaled by population size.

**5. Estimation of $\theta$.** If there is no knowledge of the distribution of allele proportions over subpopulations, moment estimators for $\theta$ may be used. There are three sums of squares that use sample allele and homozygote proportions $\tilde{p}_{a_s}$ and $\tilde{P}_{aa_s}$ in the $s$th subpopulation:

$$\text{SSP} = 2\sum_s n_s(\tilde{p}_{a_s} - \bar{p}_a)^2$$

$$\text{SSI} = \sum_s n_s(\tilde{p}_{a_s} + \tilde{P}_{aa_s} - 2\tilde{p}^2_{a_s})$$

$$\text{SSG} = \sum_s n_s(\tilde{p}_{a_s} - \tilde{P}_{aa_s})$$

where $n_s$ is the number of individuals in the $s$th sample and $\bar{p}_a$ is the mean proportion of allele $a$ over subpopulations.

If the three sums of squares are divided by d.f. $[(r-1), \sum_s (n_s - 1), \sum_s n_s]$ for $r$ samples, the resulting mean squares have expectations (over samples and subpopulations) of

$$\mathcal{E}(\text{MSP}) = \sigma^2_G + 2\sigma^2_I + 2n_c\sigma^2_P$$
$$\mathcal{E}(\text{MSG}) = \sigma^2_G + 2\sigma^2_I$$
$$\mathcal{E}(\text{MSI}) = \sigma^2_G$$

where

$$n_c = \left(\sum_s n_s - \sum_s n_s^2 / \sum_s n_s\right)/(r-1)$$
$$\sigma^2_P = \theta p_a(1-p_a)$$
$$\sigma^2_I = (F-\theta)p_a(1-p_a)$$
$$\sigma^2_G = (1-F)p_a(1-p_a)$$

so that $\theta, F$, and $p_a(1-p_a)$ can be estimated. The within-population inbreeding coefficient can be recovered from $f = (F-\theta)/(1-\theta)$ (Cockerham, 1969).

If there is independence of alleles within subpopulations, $f = 0$, and the analysis can be simplified to treat variation for alleles within (SSA) and between subpopulations (SSP) only:

$$\text{SSP} = \sum_s n_s(\tilde{p}_{a_s} - \bar{p}_a)^2$$

$$\text{SSA} = \sum_s n_s\tilde{p}_{a_s}(1-\tilde{p}_{a_s})$$

where $n_s$ is the number of alleles in the $s$th sample. If the sums of squares are divided by d.f. $[(r-1), \sum_s(n_s - 1)]$ for $r$ samples, the mean squares have expectations (over samples and subpopulations) of

$$\mathcal{E}(\text{MSP}) = \sigma_A^2 + n_c\sigma_P^2$$
$$\mathcal{E}(\text{MSA}) = \sigma_A^2$$

where

$$\sigma_P^2 = \theta p_a(1-p_a)$$
$$\sigma_A^2 = (1-\theta)p_a(1-p_a)$$

so that $\theta$ and $p_a(1-p_a)$ can be estimated.

A survey of datasets collected by forensic agencies has been conducted by B.S. Weir et al. (in preparation) and used to estimate $\theta$ between pairs of populations. Clustering populations on the basis of these genetic distances provides the same kind of clustering found for other genetic data sets (Cavalli-Sforza et al., 1994) and is consistent with current understanding of human evolution. This consistency argues for the robustness of the sampling strategies employed for the various forensic datasets. However, it is necessary to note that the quantities being estimated have interpretations that depend on the evolutionary separation of the various populations. This aspect is taken up in the next section.

**6. Hierarchical analysis.** Suppose data are available from populations for which there is a hierarchical structure, corresponding to modern humans having undergone a number of divergences starting with the African/non-African split some 100,000-200,000 years ago (Cavalli-Sforza et al., 1994). In particular, suppose data are available from $t$ sub-subpopulations within each of $s$ subpopulations within each of $r$ populations. Suppose mating is at random so that $F = \theta$ within sub-subpopulations, so the analysis can be conducted on alleles, rather than on genotypes. Then Table 5.6 of Weir (1996) has the structure

| Source | d.f. | E(MS) |
|---|---|---|
| Popns. | $r-1$ | $\sigma_A^2 + n\sigma_{SS}^2 + nt\sigma_S^2 + nst\sigma_P^2$ |
| Subpopns. | $r(s-1)$ | $\sigma_A^2 + n\sigma_{SS}^2 + nt\sigma_S^2$ |
| Sub – subpopns. | $rs(t-1)$ | $\sigma_A^2 + n\sigma_{SS}^2$ |
| Alleles | $rst(n-1)$ | $\sigma_A^2$ |

where, if $Q = p_a(1-p_a)$:

$$\sigma_A^2 = Q(1-\theta_{SS})$$
$$\sigma_{SS}^2 = Q(\theta_{SS}-\theta_S)$$
$$\sigma_S^2 = Q(\theta_S-\theta_P)$$
$$\sigma_P^2 = Q\theta_P$$

In this formulation, $\theta_{SS}$ is the relationship between pairs of alleles within the same sub-subpopulation; $\theta_S$ is the relationship of pairs of alleles within different sub-subpopulations of the same subpopulation; and $\theta_P$ is the relationship of pairs of alleles in different sub-subpopulations of different subpopulations of the same population. All are relative to the relationship of pairs of alleles from different populations.

If populations diverged $t_P$ generations ago from a population in which $\theta$ was zero; subpopulations within populations diverged $t_S$ generations ago; and sub-subpopulations within subpopulations diverged $t_{SS}$ generations ago, and if all sub-subpopulations have the same size $N$ for all time, then Equation 1 provides

$$\begin{aligned} \theta_P &= 1 - (1 - 1/2N)^{t_P - t_S} \approx (t_P - t_S)/2N < \theta_S \\ \theta_S &= 1 - (1 - 1/2N)^{t_P - t_{SS}} \approx (t_P - t_{SS})/2N < \theta_{SS} \\ \theta_{SS} &= 1 - (1 - 1/2N)^{t_P} \approx t_P/2N \end{aligned}$$

If data are available at all levels of the hierarchy, the four variance components can estimated. These allow $Q$ to be eliminated and all three coancestries $\theta_P, \theta_S, \theta_{SS}$ to be estimated.

**6.1. Data from sub-subpopulations only.** When alleles are identified only as being from the same or different sub- subpopulations, without regard to other classification, the anova structure reduces to

| *Source* | d.*f*. | E(MS) |
|---|---|---|
| *Subsubpopns.* | $rst - 1$ | $\sigma_A^2 + n[\sigma_{SS}^2 + t(rs-1)/\sigma_S^2(rst-1) + st(r-1)\sigma_P^2/(rst-1)]$ |
| *Alleles* | $rst(n-1)$ | $\sigma_A^2$ |

so that $Q$ can be eliminated and an estimate found of $\beta$:

$$\beta = \frac{\sigma_{SS}^2 + \frac{t(rs-1)}{rst-1}\sigma_S^2 + \frac{st(r-1)}{rst-1}\sigma_P^2}{\sigma_A^2 + \sigma_{SS}^2 + \frac{t(rs-1)}{rst-1}\sigma_S^2 + \frac{st(r-1)}{rst-1}\sigma_P^2}$$

For two sub-subpopulations within the same subpopulation, $r = 1, s = 1, t = 2$:

$$\begin{aligned} \beta &= \frac{\sigma_{SS}^2}{\sigma_A^2 + \sigma_{SS}^2} \\ &= \frac{\theta_{SS} - \theta_S}{1 - \theta_S} \\ &\propto t_{SS} \end{aligned}$$

For two sub-subpopulations from different subpopulations within the same population, $r = 1, s = 2, t = 1$:

$$\beta = \frac{\sigma_{SS}^2 + \sigma_S^2}{\sigma_A^2 + \sigma_{SS}^2 + \sigma_S^2}$$

$$= \frac{\theta_{SS} - \theta_P}{1 - \theta_P}$$
$$\propto t_S$$

For two sub-subpopulations from different subpopulations from different populations, $r = 2, s = 1, t = 1$:

$$\beta = \frac{\sigma_{SS}^2 + \sigma_S^2 + \sigma_P^2}{\sigma_A^2 + \sigma_{SS}^2 + \sigma_S^2 + \sigma_P^2}$$
$$= \theta_{SS}$$
$$\propto t_P$$

For each case, therefore, the same analysis leads to a distance measure that is proportional to the time since the two sampled sub-subpopulations last had an common ancestral population. Although the same symbol $\beta$ is used for these various distances they are different functions of the $\theta$'s.

The fact that $\beta$ serves as a distance to cluster populations in a way that agrees with known genetic divergences of these populations argues against it being a constant, and hence against the human population being in a stationary state.

**7. Discussion.** The United States National Research Council published a report in 1996 that provided conditional genotype probabilities for use in a forensic setting. These were the probabilities of a member (e.g. a perpetrator) of a subpopulation having a particular genotype when someone else (e.g. a suspect) in that subpopulation has already been seen to have that type. These probabilities were expressed in terms of the population structure parameter $\theta$. Logic would suggest that a similar approach be taken for the paternity setting, where now the conditional probability is for a paternal allele given the genotypes of the mother and the alleged father. This approach was advocated by Balding and Nichols (1995) but has not yet been implemented. Although the Balding and Nichols formulation requires the validity of the Dirichlet distribution for allele frequencies over populations, which can only be an approximation for real populations, the logic behind recognizing the evolutionary relationship among all humans is compelling. The practical consequences can be substantial when the people concerned belong to the same subpopulation.

**Acknowledgment.** This work was supported in part by NIH grant GM 45344.

REFERENCES

BALDING, D.J. AND R.A. NICHOLS, 1995. A method for quantifying differentiation between populations at multi-allelic loci and its implications for investigating identity and paternity. Genetica 96:3–12. Int. 64:125–140.

CAVALLI-SFORZA, L.L., P. MENOZZI AND A. PIAZZA, The History and Geography of Human Genes. 1994 Princeton NJ; Princeton University Press.

COCKERHAM, C.C., 1969. Variance of gene frequencies. Evolution 23:72–84.

COCKERHAM, C.C., 1971. Higher order probability functions of identity of alleles by descent. Genetics 69:235–246.

CROW, J.F. AND M. KIMURA, 1970. An Introduction to Population Genetics Theory. Burgess, Minneapolis MN.

GRIFFITHS, R.C., 1979. Exact sampling distributions from the infinite neutral alleles model. Adv. Appl. Prob. 11:326–354.

JIANG, C., 1987. Estimation of F-statistics in Subdivided Populations. Ph.D. thesis, Department of Statistics, North Carolina State University, Raleigh NC.

LI, Y-J., 1996. Characterizing the Structure of Genetic Populations. Ph.D. Thesis, Department of Statistics, North Carolina State University, Raleigh NC.

MALÉCOT, G., 1969. The Mathematics of Heredity. Freeman, San Francisco.

National Research Council. The Evaluation of Forensic DNA Evidence. Washington, DC; National Academy Press, 1996.

REYNOLDS, J., B.S. WEIR AND C.C. COCKERHAM., 1983. Estimation of the coancestry coefficient: Basis for a short-term genetic distance. Genetics 105:767–779.

WEIR, B.S., 1994. Effects of inbreeding on forensic calculations. Annual Review of Genetics 28:597–621.

WEIR, B.S., 1996. Genetic Data Analysis II. Sinauer, Sunderland, MA.

WRIGHT, S., 1921. Systems of mating. Genetics 6:111–178.

WRIGHT, S., 1951. The genetical structure of populations. Annals of Eugenics 15:323–354.

# COALESCENT THEORY AND ITS APPLICATIONS IN POPULATION GENETICS

WEN-HSIUNG LI* AND YUN-XIN FU*

**Abstract.** Coalescent theory represents the most significant progress in theoretical population genetics in the last two decades. It is now widely recognized as a cornerstone for rigorous statistical analyses of molecular data from populations. In addition, challenges from the rapidly expanding volume of molecular data inject fresh blood into the development of coalescent theory. This article reviews recent progress in coalescent theory and its applications, focusing on those that are important for interpreting polymorphism data from populations.

**1. Introduction.** Coalescent theory is a branch of theoretical population genetics that had its start in the early 1980's. This approach is perhaps the most important innovation in theoretical population genetics since the use of diffusion equations. Since the pioneering work of Kingman (1982a,b), it has been the most active topic in theoretical population genetics, and it is now widely recognized as the cornerstone for many statistical analyses of molecular population samples. Its popularity and power mainly come from three features. First, it is a sample-driven theory. Coalescent theory deals with the properties of a sample from a population rather than with the population as a whole. Since a population study usually relies on samples of individuals from that population, a theory that describes the properties of a sample is more relevant than the classical population genetics theory that describes the properties of the entire population. Second, it is a highly efficient approach. An important byproduct of coalescent theory is highly efficient algorithms for simulating population samples under various population genetics models, allowing various aspects of a model to be examined numerically. Third, coalescent theory is particularly suitable for molecular data, such as DNA sequence samples. This is because molecular data convey rich information about ancestral relationships among the individuals sampled.

Many aspects of coalescent theory have been reviewed previously ( Tavaré 1984, Takahata 1991, Hudson 1991, 1993, and Donnelly and Tavaré 1995). Our purpose here is to provide an overview of the topic with an emphasis on statistical methods based on coalescent theory. To be self-contained, we shall present a brief introduction of the theory and some well-known results that are important for statistical inferences based on the theory. The article consists of mainly two parts. The first is a mixture of introduction and review of coalescent theory, and the second is the review of statistical methods based on the theory.

**2. The coalescent process.** Consider a sample of $n$ sequences of a DNA region from a population of finite size and assume that there has been

*Human Genetics Center, SPH, University of Texas at Houston, Houston, TX 77030.

no recombination between sequences. Then the $n$ sequences are connected by a single phylogenetic tree or genealogy( e.g. Figure 1) where the root is the most recent common ancestor (MRCA) of these $n$ sequences. The history of the $n$ sequences can be viewed from two different perspectives (Figure 1). If one starts with the MRCA and looks forward, one sees that time and again one of the existing sequences splits into two and along the way mutations accumulate. That is, one sees *divergence.* On the other hand, if one starts with the sample of sequences, and trace backward in time, one sees that sequences become more and more similar until all of them merge to the MRCA. That is, one sees *coalescence.*

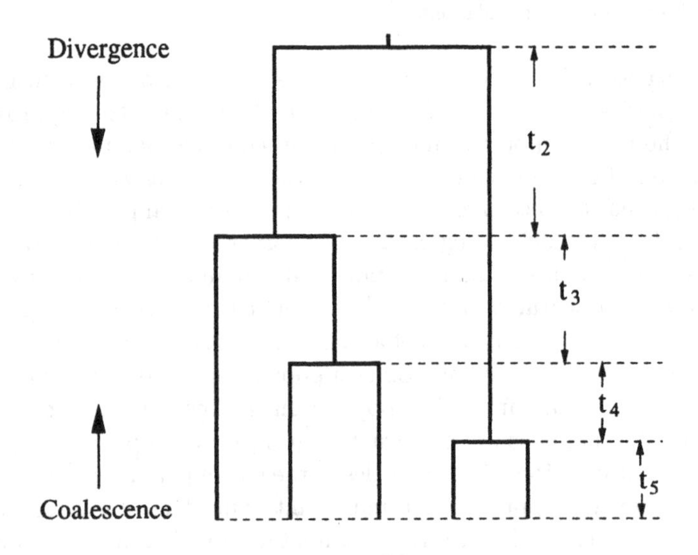


FIG. 1. *A genealogy of five sequences.*

The coalescent theory is the mathematical and statistical theory for the process of tracing sequences backward in time. It differs from the classical population genetics in two major aspects. One is that it is *retrospective* in nature, while the latter is predominantly prospective. The other is that it deals with samples from a population rather than the entire population as typically the case in classical population genetics. By dealing with samples, the theory allows rigorous analysis of experimental data, which are usually samples from populations; the retrospective nature makes the coalescent theory particularly useful for inferences about the evolution of a population from molecular data, because many questions that arise from molecular data are retrospective, as pointed out by Ewens (1979).

In general, any population study using samples and characterized by a strong retrospective component can be regarded as using the coalescent approach. Therefore, some of the well-known studies in population genetics, such as the theory of identity by decent by Cotterman (1940) and Malécot (1941), are regarded as the early work in coalescent theory (Nagylaki 1989), although Kingman (1982 a,b) are widely considered the seminal works of the field. In this section, we shall give a brief introduction of some basic components of the coalescent theory.

**Genealogical relationship among sequences:** The consequence of looking backward in time is that all the sequences in a sample will eventually coalesce to a single most recent common ancestor (MRCA), provided that the population is finite in size. The historical paths connecting individual sequences to their MRCA yield a genealogy of the sequences. When there is no selection and no population subdivision, the shape of the topological relationships among sequences is governed by a simple random process. Two sequences are randomly selected to coalesce to one at each coalescent event. If one looks top-down from the MRCA, the genealogy is a tree grown as a random bifurcation process. That is, at each branching point, a lineage is randomly selected to split into two.

Random trees such as those generated by the coalescent process also arise in some other branches of science. Harding (1971) derived a number of results about such trees. In general, many aspects of such trees can be studied using the well-known polya urn model in probability (Feller 1968). Tajima (1983) showed that the probability of having a particular topological relationship among $n$ sequences is

$$P = \frac{2^{n-1-s}}{(n-1)!}\ , \tag{1}$$

where $s$ is the number of branching points that lead to exactly two descendant sequences in the sample. For example, there are two such branching points in Figure 1, so the probability of having this tree is $2^2/4! = 1/6$.

**Distribution of coalescent times:** The time duration ($t_n$) required for $n$ sequences to coalesce to $n-1$ sequences is commonly referred to as the $n$-th coalescent time. The distributions of coalescent times are fundamental in coalescent theory because they are needed for determining the amount and pattern of DNA polymorphism in a sample. The general distribution of $t_k$ can be derived as follows:

Designate the population from which the sample was taken as generation 0 and look backward in time so that generation $i$ represents the one that was $i$ generations earlier than generation 0. For a finite population, there is a non-zero probability $p_n(i)$ that two of the $n$ sequences at generation $i$ came from one ancestral sequence at generation $i+1$. The probability that the $n$ sequences at generation $i$ coalesce to $n-1$ sequences at generation $i+t$ is therefore

$$P(t_n = t) = [1 - p_n(i)][1 - p_n(i+1)]\cdots[1 - p_n(i+t-1)]p_n(t)\ , \tag{2}$$

which is the distribution of coalescent time $t_n$. For the k-th coalescent time, we have

$$(3) \qquad P(t_k = t|s_k) = [1 - p_k(s_k)] \cdots [1 - p_k(s_k + t - 1)] p_k(s_k + t) \ ,$$

where $s_k = t_n + ... + t_{k+1}$ with $s_n = 0$. The reason why $t_k$ is dependent on $s_k$ is that the period of $t_k$ starts only when the $n$ sequences coalesce to $k$ ancestral sequences.

A simple and widely used model in population genetics is the Wright-Fisher model, which assumes that each individual contributes a very large number of gametes to the gamete pool of a population, and each individual (diploid) of the next generation is a random sample of two gametes from this gamete pool.

Consider a diploid population in which the population size ($N$) is constant over generations, that is, there are $2N$ sequences in each generation. Assume that the population evolves according to the Wright-Fisher model. Then two randomly selected sequences from the current generation have probability $1/(2N)$ of coming from the same sequence of the previous generation. Therefore,

$$p_2(i) = \frac{1}{2N} \ .$$

In general, the probability that a random sample of $k$ sequences came from $k$ different ancestral sequences of the previous generation is

$$\begin{aligned} 1 - p_k(i) &= \prod_{i=1}^{k-1} \left(1 - \frac{i}{2N}\right) \\ &\approx 1 - \frac{k(k-1)}{4N} \ , \end{aligned}$$

provided that $k$ is much smaller than $N$. Since $p_k(i)$ is independent of $i$, it follows from (3) that

$$(4) \qquad \begin{aligned} P(t_k = t) &= (1 - p_k(t))^t p_k(t) \\ &\approx \frac{k(k-1)}{4N} e^{-\frac{k(k-1)}{4N} t} \ , \end{aligned}$$

which is an exponential distribution. So we have

$$(5) \qquad E(t_k) = \frac{4N}{k(k-1)} \ ,$$

$$(6) \qquad Var(t_k) = E^2(t_k) \ .$$

Therefore, the average length of coalescent time decreases with $k$. This feature can be understood intuitively because a larger $k$ means more pairs of sequences, which means a larger chance that one of the pairs coalesces in one generation, resulting on average a shorter coalescent time.

When population size is not constant, the mathematics become complicated. Let $N_t$ be the effective population size at generation $t$. We then have from (3) that

$$p(t_k = t|s_k) = \frac{k(k-1)}{4N_{s_k+t+1}} \prod_{i=s_k+1}^{s_k+t} \left(1 - \frac{k(k-1)}{4N_i}\right) \tag{7}$$

$$\approx \frac{k(k-1)}{4N_{s_k+t+1}} exp\left[-\frac{k(k-1)}{4} \sum_{i=s_k+1}^{s_k+t} \frac{1}{N_i}\right] . \tag{8}$$

Let $v(t) = N_0/N_t$ and scale the time so that one unit correspond to $2N_0$ generations. Then a continuous approximation of the above equation results in the density function of $t_k$ as

$$f(t_k = t|s_k) \approx \frac{k(k-1)}{2} v(s_k + t) exp\left[-\frac{k(k-1)}{2} \int_{s_k}^{s_k+t} v(s)ds\right] , \tag{9}$$

which was derived by Griffiths and Tavaré (1994a).

**Time to the next event.** Coalescences are not the only type of event observable when looking backward in time. Mutations, for example, can occur in the process. Assume that the number of mutations in a given time period is a Poisson variable. Then mutations can be superimposed onto the sample genealogy, created by the coalescent process, but it is sometimes useful to consider the mutation process, as well as other types of event, together with the coalescent process. We will examine here the situation in which both the coalescent events and the mutational events are considered at the same time.

Let $\mu$ be the mutation rate per sequence per generation. Suppose that there are currently $k$ ancestral sequences. Then the probability that none of the $k$ sequences has mutations when transmitted to the current generation from the previous one is

$$(1-\mu)^k \approx 1 - k\mu .$$

The probability that a mutation occurs $t+1$ generations earlier is thus

$$k\mu(1-k\mu)^t \approx k\mu e^{-k\mu t} ,$$

which is also an exponential distribution.

The coalescent process and the mutation process competes against each other for the next event. Assume a constant population size. At each generation the probability that either a coalescent or mutational event occurs is $k(k-1)/(4N) + k\mu = \frac{1}{4N}[k(k-1)+k\theta]$ where $\theta = 4N\mu$. Therefore the probability that the next event is a coalescent is

$$\frac{k-1}{k-1+\theta} \tag{10}$$

and the probability that the next event is a mutation is

$$\frac{\theta}{k-1+\theta} . \tag{11}$$

By the same argument for the coalescent time distribution, the time length to the next event is therefore exponentially distributed with mean equal to

$$\frac{4N}{k(k-1)+k\theta} .$$

**Mutations in the genealogy of a sample:** Assume that the number of mutations occurring in a sequence in a given period of time $T$ is a Poisson variable with mean equal to $\mu T$. Knowing the distributions of coalescent times allows one to quantify various aspects of the mutations in a sample. We will consider a few cases under the neutral Wright-Fisher model with a constant effective population size.

The simplest quantity is the number $K$ of mutations in a sample of size $n$ since the MRCA. Let $K_i$ be the number of mutations during the period $t_i$, so that $K = K_2 + ... + K_n$. It is simple to see that

$$E(K|t_2, ..., t_n) = \mu(2t_2 + ... + nt_n) ,$$

because $K$ conditioning on the $t_i's$ is the sum of $n-1$ independent Poisson variables with means equal to $\mu 2t, ..., \mu n t_n$, respectively. Therefore, the expectation of $K$ is

$$E(K) = E_{t_2...t_n} E(K|t_2...t_n) = \mu E(\sum_{k=2}^{n} kt_k) \tag{12}$$

$$= a_n \theta , \tag{13}$$

where $E_{t_2...t_n}$ stands for expectation with respect to $t_2, ..., t_n$, and

$$a_n = 1 + \frac{1}{2} + \cdots + \frac{1}{n-1} . \tag{14}$$

The variance of $K$ is

$$\begin{aligned} Var(K) &= E(K^2) - E^2(K) \\ &= E_{t_i's} E(K^2|t_2, ..., t_n) - 4\mu^2 E^2(t_2) \\ &= a_n\theta + b_n\theta^2 \end{aligned} \tag{15}$$

where

$$b_n = 1 + \frac{1}{4} + \cdots + \frac{1}{(n-1)^2} . \tag{16}$$

The mean number $\Pi$ of nucleotide differences between two sequences is another commonly used quantity. Let $d_{ij}$ be the number of nucleotide differences between sequences $i$ and $j$. Then $\Pi$ is defined as

$$\Pi = \frac{2}{n(n-1)} \sum_{i<j} d_{ij} \; . \tag{17}$$

Since for a random pair of sequences, $E(d_{ij}) = \theta$, it follows that

$$E(\Pi) = \theta = 4N\mu \; . \tag{18}$$

The variance of $\theta$ is more complicated, but was derived by Tajima (1983) as

$$Var(\Pi) = \frac{n+1}{3(n-1)}\theta + \frac{2(n^2+n+3)}{9n(n-1)}\theta^2 \; . \tag{19}$$

Mutations in a genealogy can be further partitioned into different categories. The genealogy of a sample of $n$ sequences consists of $2(n-1)$ branches and each branch has at least one sequence in the sample as its descendant. Define the number of sequences in a sample that are descendants of a branch as the **size** of that branch. A mutation is said to be **size** $i$ if it occurs on a branch of size $i$ (Figure 2). It is easy to see that a mutation of size 1 can only occur in an external branch, i.e., a branch that directly connects to an external node (sequence). For this reason, a mutation of size 1 is often referred to as an **external mutation**.

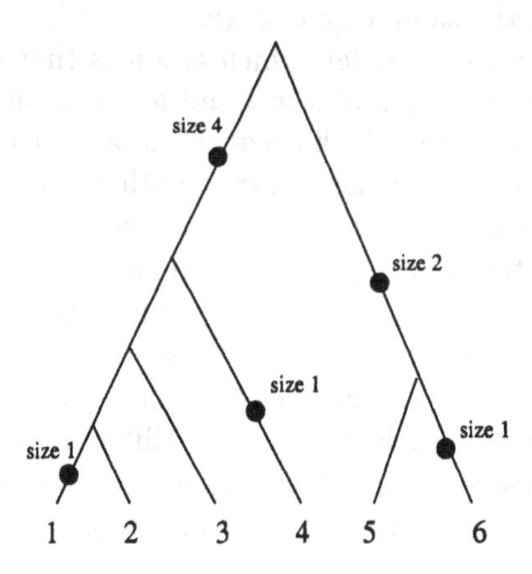


FIG. 2. *A genealogy of six sequences with five mutations, three of which are of size 1 (external mutations), one of size 2 and one of size 4; or three mutations of type 1 and two of type 2 (see text for definitions).*

Let $\xi_i$ be the number of mutations of size $i$. Fu and Li (1993a) showed that $E(\xi_1) = \theta = 4N\mu$, so that the expected number of external mutations does not depend on the sample size. Fu (1995) showed that

$$E(\xi_i) = \frac{1}{i}\theta ,$$

$$Var(\xi_i) = \frac{1}{i}\theta + \sigma_{ii}\theta^2 , \tag{20}$$

where $\sigma_{ii}$ is given by

$$\sigma_{ii} = \begin{cases} \beta_n(i+1), & \text{if } i < \frac{n}{2}; \\ 2\frac{a_n - a_i}{n-i} - \frac{1}{i^2}, & \text{if } i = \frac{n}{2}; \\ \beta_n(i) - \frac{1}{i^2}, & \text{if } i > \frac{n}{2}. \end{cases} \tag{21}$$

where

$$\beta_n(i) = \frac{2n}{(n-i+1)(n-i)}(a_{n+1} - a_i) - \frac{2}{n-i} . \tag{22}$$

The covariance between $\xi_i$ and $\xi_j$, which is in the form $\sigma_{ij}\theta^2$, is also given by Fu (1995).

Since $K_n = \xi_1 + ... + \xi_{n-1}$, $E(K_n) = E(\xi_1) + ... + E(\xi_{n-1}) = \theta(1 + 1/2 + ... + 1/(n-1))$. This is equation (13) and has two different interpretations. The first is that on average $\theta/i$ mutations occur during coalescent time period $t_i$. The second is that on average $\theta/i$ mutations occur on the branches of size $i$ in the sample genealogy.

Under the infinite-site model, which assumes that each mutation occurs at a new site, a mutation of size $i$ and a mutation of size $n-i$ both result in a segregating site at which $i$ sequences carry one type of nucleotide and the remaining $n-i$ sequences carry another type of nucleotide. We define a mutation as **type** $i$ if it is either of size $i$ or of size $n-i$. When the ancestral nucleotide of a segregating site is unknown, one cannot distinguish between a mutation of type $i$ and a mutation of type $n-i$ without additional analysis. For this reason, it is convenient to group mutations into types rather than sizes. Since a type 1 mutation is characterized by one sequence carrying a nucleotide that is different from that of the remaining $n-1$ sequences, it is often referred to as a **singleton**. Define $\eta_i$ $(i = 1, ..., n/2)$ to be the number of mutations of type $i$. Then the expectation of $\eta_i$ is

$$E(\eta_i) = \begin{cases} \frac{\theta}{i} + \frac{\theta}{n-i}, & \text{if } i \neq n-i , \\ \frac{\theta}{i}, & \text{if } i = n-i . \end{cases} \tag{23}$$

The variance of $\eta_i$ and the covariance of $\eta_i$ and $\eta_j$ can be calculated from those of $\xi$'s. The details can be found in Fu (1995).

**Probability of observing a polymorphism pattern in a sample.** We presented means and variances of a number of summary statistics in the previous sections. The moments of these statistics are very useful in practice because they are the basis for some fast statistical analyses of a sample. However, for certain complex problems more powerful approaches are necessary. Being able to compute the probability of a polymorphism pattern provides the necessary tool for powerful statistical methods such as the maximum likelihood and Bayesian methods.

The probability of $K$ segregating sites in a sample of $n$ sequences is a convolution of $n-1$ geometric distributions (Watterson 1975) and its explicit form is as follows (Tavaré 1984):

$$p(K) = \frac{n-1}{\theta} \sum_{i=0}^{n-2} (-1)^i \binom{n-2}{i} \left( \frac{\theta}{\theta + i + 1} \right)^{K+1}. \tag{24}$$

When the polymorphism in a sample is examined sequence by sequence, one arrives at a set of haplotypes and numbers of mutations separating them. Strobeck (1983) appears to be the first to use recurrence equations to study the probability of haplotypes from DNA sequences, but he considered at most only three alleles. Ether and Griffiths (1987) and Griffiths (1989) substantially advanced this direction of research. To date, many recurrence equations of probabilities of various forms of sample polymorphism have been derived (Sawyer et al. 1987, Griffiths and Tavaré 1994c, 1995, Lundstrom et al. 1992, Krone and Neuhauser 1997,Neuhauser and Krone 1997, and Fu 1997c, 1998). We shall use an example to illustrate one of the recurrence equations in Griffiths and Tavaré (1995).

Consider the following example of six sequences resulting from the five mutations in Fig. 2:

$$\begin{array}{c}
1 \cdots 0 \cdots 1 \cdots 0 \cdots 0 \\
0 \cdots 0 \cdots 1 \cdots 0 \cdots 0 \\
0 \cdots 0 \cdots 1 \cdots 0 \cdots 0 \\
0 \cdots 1 \cdots 1 \cdots 0 \cdots 0 \\
0 \cdots 0 \cdots 0 \cdots 1 \cdots 0 \\
0 \cdots 0 \cdots 0 \cdots 1 \cdots 1
\end{array}$$

where 0 and 1 represent the ancestral and mutant nucleotides, respectively, and dots represent the intervening sequence segments between segregating sites. Denote this polymorphism pattern as $S$, which is a $6 \times 5$ matrix with rows representing sequences (not necessarily distinct) and columns representing mutations.

Assume that this sample was taken from a random mating population with a constant $\theta$. To compute the probability $P(S)$ of observing $S$, we can start with six sequences and follow the coalescent process and the mutation process backward in time. The next event is either a coalescent event or a mutational event.

Consider the coalescent event first. Among all possible coalescent pairs only the coalescence by sequences 2 and 3, resulting in polymorphism pattern $S_2$, has the potential of leading to a sample genealogy that is compatible with $S$, where the subscript 2 means removal of the second sequence in the sample (it is the same as removing sequence 3). The probability that the coalescence is between sequences 2 and 3 is $2/(5 \times 6)$.

Suppose the next event is a mutation instead of a coalescence. Then the mutation can happen with equal probability in any of the six sequences, but the consequence is potentially compatible with the polymorphism pattern $S$ only when it occurs on sequences $1, 4$ or 6. Take sequence 1 for example. A mutation on this sequence must result in polymorphism $S^1$ for it to be compatible with $S$, where superscript 1 means removal of the first segregating site, which corresponds to the mutation having occurred on sequence 1.

It follows from the above analysis and Equations (10) and (11) that we can write $P(S)$ as the following recurrence equation:

$$
\begin{aligned}
P(S) = & \left(\frac{5}{5+\theta}\right)\left(\frac{2}{5 \times 6}\right) P(S_2) \\
& + \left(\frac{\theta}{5+\theta}\right)\left(\frac{1}{6}\right)\left[P(S^1) + P(S^2) + P(S^5)\right] .
\end{aligned}
$$

This recurrence equation implies that to compute $P(S)$, we only need to know $P(S_2), P(S^2)$ and $P(S^5)$, which can in turn be expressed by their respective recurrence equations. Expressing $P(S)$ by recurrence equations allows one to compute its value sequentially from those of smaller sample sizes and fewer mutations.

The above analysis can be generalized. Let $\boldsymbol{S}$ be the matrix of distinct haplotypes (in the form of a matrix with elements being either 0 or 1) and $\boldsymbol{n}$ be the vector of their multiplicities. For instance, for the above example, $\boldsymbol{S}$ is the matrix obtained by removing the second row of $S$, and $\boldsymbol{n} = (1, 2, 1, 1, 1)$. Griffiths and Tavaré (1995) showed that the probability

$P(\boldsymbol{S}, \boldsymbol{n})$ of sample configuration $(\boldsymbol{S}, \boldsymbol{n})$ satisfies the recurrence equation:

$$\begin{aligned} n(n-1+\theta)P(\boldsymbol{S},\boldsymbol{n}) = & \sum_{k:n_k>1} n_k(n_k-1)P(\boldsymbol{S},\boldsymbol{n}-\boldsymbol{e}_k) \\ & + \theta \sum_{k:n_k=1,\boldsymbol{s}_{.l}=\boldsymbol{e}_k,\boldsymbol{s}^l_{k.}\neq\boldsymbol{s}^l_{j.},k\neq j} P(\boldsymbol{S}^l,\boldsymbol{n}) \\ & + \theta \sum_{k:n_k=1,\boldsymbol{s}_{.l}=\boldsymbol{e}_k,\boldsymbol{s}^l_{k.}=\boldsymbol{s}^l_{j.}} P(\boldsymbol{S}^{kl},(\boldsymbol{n}+\boldsymbol{e}_j)^k) , \end{aligned} \tag{25}$$

where $\boldsymbol{e}_k$ is the k-th unit row vector, i.e., the k-th element is 1 and all the remaining $k-1$ elements are 0, $\boldsymbol{s}_{i.}$ and $\boldsymbol{s}_{.l}$ denote the $i$-th row and the transpose of the $l$-th column of $\boldsymbol{S}$. A superscript denotes removal of that column, and a double superscript notation $S^{kl}$ indicates removal of row $k$ and column $l$. The complicated condition for the second summation simply means that the summation is taken over those external mutations such that the removal of such an external mutation does not reduce the number of distinct haplotypes. The third summation is taken over those external mutations such that the removal of such an external mutation results in reducing the number of distinct haplotypes by one.

Although the collection of haplotypes and mutations separating them captures all the polymorphism information in a sample, they are not always the best representation of data for a given inference problem. From the experimental point of view, haplotypes are more costly to obtain than a pattern of segregating sites for a large autosomal region using the PCR (Polymerase Chain Reaction) technique, because the latter does not require distinction of the two alleles in an individual. In humans, hundreds of polymorphic sites in the nuclear genome have been identified (e.g. Wang et al. 1996) and many more will be available with the progress of the Human Genome Project. Using hundreds or even thousands of known polymorphic sites to screen a population sample would generate a huge amount of information about the history of that population in the form of a pattern of segregating sites. Therefore, it would be very useful to develop methods for computing the probability of a segregating pattern in a sample. Fu (1997c) derived a number of recurrence equations for the probabilities of $\boldsymbol{\xi}$ and $\boldsymbol{\eta}$, and their marginal probabilities, such as the probability of $k$ external mutations. Fu's recurrence equations are different from those for haplotypes but recurrence equations for these probabilities analogous to (25) have also been derived (Fu 1998).

**Coalescent in complex models.** Coalescent theory has been developed for a number of more complex population genetics models. Common to these developments is that knowledge of coalescent events is not enough to fully characterize the process, and other events must be considered together with the coalescent events. For example, when recombination is considered, the sequences involved in recombination and the places where recombinations occur need to be tracked as well. Hudson (1991) gave a comprehensive review on coalescent theory with recombination, population division and certain types of selection. Incorporating selection into coalescent analysis proves to be a challenging task and the mathematics becomes quite involved. For recent work involving natural selection, see Hudson and Kaplan (1994) and Krone and Neuhauser (1997).

Another active line of research is the coalescent theory for non-randomly mating population, such as plant populations, many of which undergo partial self-fertilization. As plant population geneticists are beginning to utilize $DNA$ sequence data (Clegg 1996), coalescent theory for non-randomly mating populations will be increasingly important. This line of research have been initiated by Slatkin (1991), who attempted to connect the inbreeding coefficients to the coalescent theory. Milligan (1996) stimulated further development by proposing an estimator of selfing rate and an estimator of $\theta$ using Slatkin's approach. The coalescent theory for partially selfing populations parallel to the randomly-mating population has been recently developed by Fu (1997a) and Nordberg and Donnelly (1997).

**3. Applications.** If one regards characterizing the coalescent process under various models and properties of polymorphisms in a sample as the theory of coalescent, then inference methods based on the theory as well as analyses of population samples using these methods can be considered as its applications. Statistical inferences require integration of the coalescent theory and statistical principles and techniques, and they need to deal with the quality of the inferences. Furthermore, very often the development of a statistical method based on coalescent is accompanied by a development of some aspects of the coalescent theory. Therefore, applications of coalescent theory are by no means simple. In a sense, developing a proper statistical method is often harder. For example, the coalescent times are the basic ingredient of the coalescent theory and their properties under a neutral model are quite straightforward. Estimation of the coalescent times, however, is not a simple matter because coalescent times are not observable in experiment and they must be inferred from the polymorphism pattern in a sample. In this section, we shall review a number of statistical methods developed recently.

**4. Estimation of $\theta$.** The quantity $\theta = 4N\mu$ is the most important parameter for the evolution of a DNA region in a population. As described in earlier sections, many summary statistics for a DNA sample are related to $\theta$. Therefore, estimating $\theta$ with high accuracy plays an important role in understanding the evolution of a population. Two commonly used estimators of $\theta$ are due to Watterson (1975) and Tajima (1983). Watterson's estimator is defined as

$$\theta_W = K/a_n \,, \tag{26}$$

where $K$ is the number of segregating sites in the sample and $a_n$ is given by (14). Tajima's estimator is based on (18) and is defined as

$$\theta_T = \Pi \,. \tag{27}$$

Both estimators are unbiased. Since Watterson's estimator has smaller variance than Tajima's, which can be seen from Equations (15) and (19), $\theta_W$ is a better estimator than $\theta_T$ under neutrality.

The above two estimators make little use of the genealogical relationships among the sequences in the sample. Whether such information can lead to better estimators was discussed by Felsenstein (1992a) and Fu and Li (1993b). The reason why there was even a debate is due to the widespread belief that Watterson's estimator is the best estimator one can derive for a DNA sample, analogous to Ewens' estimator of $\theta$ based on the number of alleles in a sample. Felsenstein (1992a) considered a sample of sequences of infinite length and suggested that much improvement can be achieved by incorporating genealogical relationships. As only finite sequences are available, Fu and Li (1993a) investigated the maximum amount of improvement that might be achieved in practice.

Let us consider the hypothetical genealogy in Figure 1. The tree is divided into segments by the horizontal lines through each of the branching nodes (coalescent events). Suppose that in the ideal situation, we can observe or infer without error the number $k_{ij}$ of mutations in the $j$-th lineage during $t_i$, $(i = 2, ..., n; j = 1, ..., i)$. Fu and Li (1993a) showed that the optimal estimator $\theta_m$ is the solution of the following equation

$$\sum_{i=2}^{n} \frac{k_i + 1}{\theta + i - 1} = \frac{K}{\theta} \,, \tag{28}$$

where $k_i = \sum_j k_{ij}$ is the number of mutations that occurred during the period $t_i$, and $K = \sum_{i=2}^{n} k_i$ is the total number of mutations in the genealogy. Note that this optimal estimator only requires knowledge of the $n-1$ values of $k_i$ and does not require knowledge of the explicit relationships among the $n$ sequences. The large-sample variance of the estimator is

$$V_{min} = \theta \left[ \sum_{i=2}^{n} \frac{1}{\theta + i - 1} \right]^{-1} \,. \tag{29}$$

Since $\theta_m$ is based on $k_2, ..., k_n$, whereas $\theta_W$ is based on only $\eta$, $\theta_m$ should be more efficient than $\theta_W$. The ratio $V_{min}/V(\theta_W)$ indicates the relative efficiency of $\theta_W$ to $\theta_m$. Figure 3 shows that this ratio is much smaller than one, if $\theta$ is large. Thus, $\theta_m$ can be much more efficient than $\theta_W$. Although it is unlikely that $\eta_i$ can be inferred without error, this analysis does suggest that there is considerable room for improvement over Watterson's estimator $\theta_W$. For this reason, there has been much effort to improve the estimation

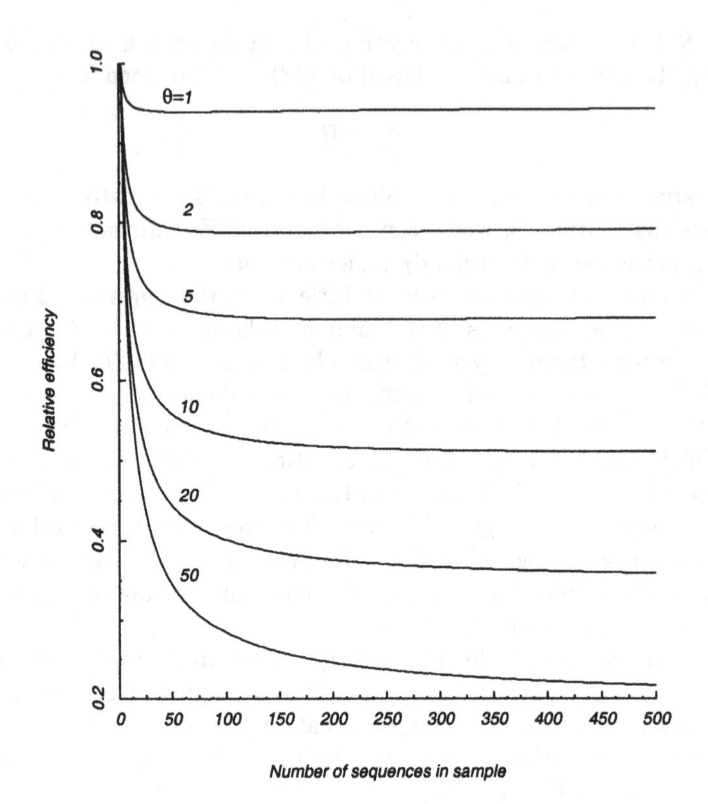


FIG. 3. *Efficiency of Watterson's estimate $\theta_W$ of $\theta$. The six curves from top to bottom correspond, respectively, to $\theta = 1, 2, 5, 10, 20$ and 50. The efficiency is calculated by $V_{min}/V(\theta_W)$. From Fu and Li (1993a).*

of $\theta$. We will describe the new methods of Fu (1994ab), Griffiths and Tavaré (1994c, 1995), Kuhner et al. (1995), and Fu (1997c).

*BLUE estimators:* Fu and Li's (1993a) analysis shows that a better estimator of $\theta$ can be found when one partitions the total number $K$ of mutations into some non-overlapping categories and derives an estimator based on the more detailed information. There are a number of ways to partition $K$. We have seen in the above that $K$ can be partitioned into $n-1$ groups corresponding to coalescent times. Since it is unlikely that $k_i$ can be inferred with high accuracy in practice, we need some alternatives. Fu (1994a) partitioned $K$ according to on which branch the mutations occurred. Since there are $2(n-1)$ branches in the sample genealogy (see Fig. 1), $K$ mutations are partitioned into $2(n-1)$ categories. Let $m_i$ be the number of mutations occurring on branch $i$. Fu (1994a) shows that

$$E(m_i) = \alpha_i\theta \ , \tag{30}$$

$$Var(m_i) = \alpha_i\theta + \beta_{ii}\theta^2 \ , \tag{31}$$

$$Cov(m_i, m_j) = \beta_{ij}\theta^2 \ , \tag{32}$$

where $\alpha_i, \beta_{ij}$ are constants that are determined by the topology of the genealogy of the sample. A simple example of the values of $\alpha_i, \beta_{ij}$ is given in Li and Fu (1994). Define $\boldsymbol{m} = (m_1, ..., m_{2(n-1)})^T$ and $\boldsymbol{\alpha} = (\alpha_1, ..., \alpha_{2(n-1)})^T$. Then we have the following linear model:

$$\boldsymbol{m} = \boldsymbol{\alpha}\theta + \boldsymbol{\epsilon} \ , \tag{33}$$

where the error term $\boldsymbol{\epsilon} = \boldsymbol{m} - \boldsymbol{\alpha}\theta$ and

$$Var(\epsilon) = \boldsymbol{\gamma}\theta + \boldsymbol{\beta}\theta^2 \ , \tag{34}$$

where

$$\boldsymbol{\gamma} = diag(\alpha_1, ..., \alpha_{2(n-1)}) \ , \tag{35}$$

$$\boldsymbol{\beta} = \{\beta_{ij} : i, j = 1, ..., 2(n-1)\} \ . \tag{36}$$

The best linear unbiased estimator (BLUE) of $\theta$ can be found as

$$\theta = \left[\frac{\boldsymbol{\alpha}^T(\boldsymbol{\gamma} + \theta\boldsymbol{\beta})^{-1}}{\boldsymbol{\alpha}^T(\boldsymbol{\gamma} + \theta\boldsymbol{\beta})^{-1}\boldsymbol{\alpha}}\right]\boldsymbol{m} \ . \tag{37}$$

However, this equation does not give us an estimate of $\theta$ directly because the computation of $\boldsymbol{\gamma}+\theta\boldsymbol{\beta}$ requires the value of $\theta$ which is unknown. Therefore, an iteration procedure is necessary. Obviously, an estimate of $\theta$ can be obtained as the limit of the series

$$\theta_{k+1} = \left[\frac{\boldsymbol{\alpha}^T(\boldsymbol{\gamma} + \theta_k\boldsymbol{\beta})^{-1}}{\boldsymbol{\alpha}^T(\boldsymbol{\gamma} + \theta_k\boldsymbol{\beta})^{-1}\boldsymbol{\alpha}}\right]\boldsymbol{m} \ , \tag{38}$$

taking $\theta_0$ an arbitrary non-negative number.

To use BLUE as an estimate of $\theta$, the sample genealogy must be inferred. Because an inferred genealogy often contains errors, the BLUE of $\theta$ based on an erroneous genealogy is likely to be biased. It is therefore important in practice to be able to correct the estimation bias. Under the neutral Wright-Fisher model, the unweighted pair-group method with arithmetic mean (UPGMA), the simplest tree-making method, seems to be adequate for reconstructing the sample genealogy( see Nei 1987 or Li 1997 for description of the method). To use UPGMA, one needs to calculate the number of mutations separating each pair of sequences (under the infinite sites model, this number is the same as the number of nucleotide differences between the two sequences). These numbers form a distance matrix upon which the UPGMA is applied to obtain the genealogy of the sample. Fu (1994a) found that when the BLUE procedure is applied to the genealogy reconstructed by UPGMA, there is indeed, on average, a downward bias in the estimate of $\theta$. The bias can be corrected using the following regression equation

$$\hat{\theta} = \left(0.0335\sqrt{n-2} + 0.997\sqrt{\hat{\theta}_U}\right)^2, \tag{39}$$

where $\hat{\theta}_U$ is the BLUE estimate of $\theta$ based on the genealogy reconstructed by UPGMA. Fu (1994a) called this estimator UPBLUE and found that UPBLUE is nearly unbiased and has a variance close to the minimum variance $V_{min}$. In other words, UPBLUE is a nearly optimal estimator of $\theta$ – a demonstration that a practical estimator of $\theta$ can be substantially better than Watterson's estimator $\theta_W$. Deng and Fu (1996) studied the effect of variable mutation rate among nucleotides on UPBLUE and found that UPBLUE is quite robust against rate heterogeneity when the sample size is reasonably large. They also gave a new regression equation for bias correction when it is warranted.

It is obvious from our earlier discussion on various types of mutations that $\boldsymbol{\xi}$ and $\boldsymbol{\eta}$ can be used for deriving estimators of $\theta$ similar to the BLUE. Indeed, if we replace in Equation (33) $\boldsymbol{m}$ by $\boldsymbol{\xi}$, and defined $\boldsymbol{\alpha}$ and $\boldsymbol{\beta}$ accordingly, BLUE estimator (38) can be applied. Fu (1994b) showed that the BLUE estimator based on $\boldsymbol{\xi}$ is also very efficient, having a variance only marginally larger than that of UPBLUE. An advantage of BLUE based on $\boldsymbol{\xi}$ is that no correction of bias is needed, Another advantage is that it can easily be adopted to more complex population genetics models. For example, Fu (1994b) used this estimator in models with subdivided populations, and with recombination.

*Maximum likelihood estimators:* To date, three different maximum likelihood methods have been developed for estimating $\theta$. We shall describe them in turn.

One method by Griffith and Tavaré (1994c, 1995) is based on Markov chain Monte-Carlo to evaluate the probability of a sample configuration that is in the form of recurrence equation such as (25). The idea is to construct a Markov chain so that the probability can be expressed as a function of the Markov chain and can be estimated by multiple realizations of the Markov chain. To illustrate the approach, consider $P(\boldsymbol{S}, \boldsymbol{n})$ in equation (25). Looking backward in time, there are a number of configurations that are compatible with $(\boldsymbol{S}, \boldsymbol{n})$ when the next event occurs. Let $\{X(l), l = 0, ...\}$ be a Markov chain with transitions as follows:

$$(40)\quad (\boldsymbol{S},\boldsymbol{n}) \to (\boldsymbol{S},\boldsymbol{n}-\boldsymbol{e}_k) \text{ with probability } \frac{n_k(n_k-1)}{f(\boldsymbol{S},\boldsymbol{n})n(n-1+\theta)}$$

$$(41)\quad \to (\boldsymbol{S}^l,\boldsymbol{n}) \text{ with probability } \frac{\theta}{f(\boldsymbol{S},\boldsymbol{n})n(n-1+\theta)}$$

$$(42)\quad \to (\boldsymbol{S}^{kl},(\boldsymbol{n}+\boldsymbol{e}_j)^k) \text{ with probability } \frac{\theta}{f(\boldsymbol{S},\boldsymbol{n})n(n-1+\theta)},$$

where $f(\boldsymbol{S},\boldsymbol{n})$ is a scale so that the sum of above transition probabilities equals one. That is,

$$(43)\qquad f(\boldsymbol{S},\boldsymbol{n}) = \frac{1}{n(n-1+\theta)}\left[\sum_{k:n_k>1} n_k(n_k-1) + \xi_1\theta\right],$$

where $\xi_1$ is the number of mutations of size 1 in $(\boldsymbol{S},\boldsymbol{n})$.

Let $(\boldsymbol{S}(l),\boldsymbol{n}(l))$ be the state of the Markov chain $\{X\}$ at step $l$. Then $\{X\}$ starts at state $(\boldsymbol{S},\boldsymbol{n})$ and ends when the MRCA is reached. A remarkable lemma by Griffiths and Tavaré (1994c) is that

$$(44)\qquad P(\boldsymbol{S},\boldsymbol{n}) = E_{(\boldsymbol{S},\boldsymbol{n})}\prod_l f(\boldsymbol{S}(l),\boldsymbol{n}(l)).$$

This representation of $P(\boldsymbol{S},\boldsymbol{n})$ suggests that it can be estimated as the mean of $\prod_l f(\boldsymbol{S}(l),\boldsymbol{n}(l))$ over replicates of the Markov process. To estimate $\theta$, one can estimate $P(\boldsymbol{S},\boldsymbol{n})$ for different values of $\theta$, and the one maximizing the value of $P(\boldsymbol{S},\boldsymbol{n})$ is the maximum likelihood estimate of $\theta$. Griffiths and Tavaré (1994c, 1995) show that $P(\boldsymbol{S},\boldsymbol{n})$ for different values of $\theta$ can be estimated by those replicates with a single $\theta$. This and several other variance reduction techniques significantly reduce the amount of computation required. Griffiths and Tavaré (1994a) extended this approach to the situation in which the population grows exponentially.

A maximum likelihood estimator developed by Felsenstein and his colleagues (Felsenstein 1992b, Kuhner et al. 1995) also uses the Monte-Carlo method to evaluate the probability of a sample. They noted that the probability of the polymorphism pattern in a sample can be written as

$$(45)\qquad P(Data|\theta) = \sum_G P(Data|G)P(G|\theta),$$

where $G$ represents the genealogy of the sample, including the coalescent times. The summation is taken over all possible genealogies. $P(G|\theta)$ is easy to compute because it is the product of the probability of a topology and the probability of a given set of coalescent times. $P(Data|G)$ is the probability of the sample given genealogy $G$, and its computation is the same as that for an inter-specific phylogeny. The problem is, however, that there are too many genealogies to examine in practice. A strategy to examine genealogies selectively must be implemented to estimate $P(Data|\theta)$ using (45) as the basis. Kuhner et al. (1995) used the Metropolis-Hastings method (Metropolis et al. 1953, Hastings 1970). Start with an initial genealogy $G$. Then make small changes in the genealogy and compute $P(Data|G)$ for the new genealogy. Test the genealogy to see if it should be accepted as a new one. If it does, the new genealogy becomes the current one, then repeat the whole process. If it does not, then select another genealogy and test it as described above. This process creates a Markov chain of genealogies from which an estimate of $\theta$ can be obtained after a sufficient number of repeats. The likelihood values for different $\theta$ can also be estimated by the realizations of the Markov chain with a single $\theta$. A limited simulation by Kuhner et al. (1995) showed that this method indeed produces a better estimate than Watterson's estimator.

Inspired by an earlier work showing the efficiency of the BLUE based on $\boldsymbol{\xi}$ (Fu, 1994b), Fu (1997c) suggested another maximum likelihood method based on $P(\boldsymbol{\xi})$, taking advantage of being able to evaluate this probability using recurrence equations. At present, this method is only capable of handling a modest sample size with not too many segregating sites. A Markov chain Monte-Carlo method for evaluating $P(\boldsymbol{\xi})$ similar to those of Griffiths and Tavaré (1995) is currently being developed (Fu 1998).

Before leaving this subject, we note that estimation of $\theta$ is associated with considerable variance even with the best possible estimator. All the estimators described in this section are either nearly unbiased or expected to be so, and are expected to have a variance smaller than Watterson's estimate $\theta_W$. Therefore, they should give reasonably good estimates. From the practical point of view, it does not make much difference which of the newer methods is used. A BLUE based estimator has the advantage of being computationally fast with proven accuracy. Among the likelihood-based methods, Griffiths and Tavaré's approach is mathematically more elegant than Felsenstein's method. The latter method has the advantage of being extendible to more complex nucleotide substitution models although such extension may not be necessary in most situations. The maximum likelihood estimator based on $\boldsymbol{\xi}$ may prove to be very useful when a large autosomal region is studied and when haplotypes are not available. Finally, it is worth speculating why UPBLUE, which is based on a single estimated genealogy of a sample, can produce a nearly unbiased estimate of $\theta$ with almost the smallest variance possible. We think that the answer lies in Equation (28), which shows that to obtain the best estimate of $\theta$, all we need to know are the numbers $k_i$ of mutations in state $i$ ($i = 2, ..., n$) of the coalescent. Genealogy per se has nothing to do with the parameter $\theta$, its role is to improve our reconstruction of $k_i$. Therefore, errors in an estimated sample genealogy do not necessarily translate into error in estimation of $\theta$ as long as $k_i$ from the estimated sample genealogy do not differ much from their true values. UPGMA appears to be capable of producing a tree that is sufficiently accurate for this purpose. Consequently UPBLUE is a highly efficient estimator of $\theta$.

**5. Estimation of the recombination rate.** The recombination parameter $C = 4Nr$, where $r$ is the recombination rate per sequence per generation, is also an important parameter shaping the pattern of polymorphisms in a sample. Recombination breaks linkage between nucleotides and creates correlated multiple histories for the nucleotides of sequences in a sample. $C$ is more difficult to estimate than $\theta$, because recombinations are not always traceable from a sample, while most mutations leave their marks as segregating sites, from which $\theta$ can be estimated. $C$ must be estimated from the pattern of polymorphisms.

The so-called "Four-gamete test" by Hudson and Kaplan (1985) can be used to check whether there was recombination between two segregating sites. When there is no recombination, it is easy to verify that there can be at most three alleles in four sequences assuming the infinite-site model. Therefore, the existence of four alleles indicates at least one recombination event. Hudson and Kaplan (1985) developed an algorithm for finding the minimum number $R_m$ of recombinations in a sample. They showed that the expectation of $R_m$ is much smaller than the expected number of recombinations. An estimate of $C$ can be obtained by comparing $R_m$ to its expected value for difference values of $C$, and then choose the estimate of $C$ as the one that gives the closest match between $R_m$ and its expectation.

Hey and Wakeley (1997) extended Hudson and Kaplan's approach. Instead of considering expectations, they attempted to derive the probability $I_{2,n}$ of observing four alleles when two informative segregating sites separated by $n$ non-informative nucleotides in four sequences are examined. A segregating site is said to be informative if the mutation is not of type 1. They showed that

$$I_{2,n} \approx \frac{2}{3} - \frac{1}{6}\left(\frac{C}{C+3} - 2\right)^2 exp\left[-\frac{3C(n-1)}{5(C+3)}\right] \tag{46}$$

for $n > 0$ under the assumption that the polymorphism patterns defined by different pairs of nucleotides are independent. For three informative sites in four sequences, they obtained an estimate of $C$ as follows. Consider the product $L$ of the probabilities of allele numbers defined by 1st and 2nd segregating site, and 2nd and 3rd segregating sites. If, for example, there are four alleles by the 1st and 2nd segregating sites and two alleles by the 2nd and 3rd segregating sites, then $L = I_{2,n_1}(1 - I_{2,n_2})$ where $n_1$ and $n_2$ are the number of nucleotides separating 1st and 2nd, and 2nd and 3rd segregating sites, respectively. The value of $C$ that maximizes $L$ is taken as their estimate of $C$. The final estimate $\gamma$ of $C$ is the average of estimates of $C$ over all possible subset of four sequences and all combinations of three informative sites. Hey and Wakeley (1997) showed that their estimate $\gamma$ performs better than Hudson's when the sample size is small but about the same when the sample size is reasonably large.

In addition to the unrealistic assumption leading to (46), Hey and Wakeley's estimator appears to suffer from the same shortcoming as Tajima's estimator $\theta_T$ of $\theta$. When averaging estimates over all subsets of sequences, the portion of sample genealogy close to the root is heavily weighted, resulting in inefficient use of information and consequently large sampling variance in the estimation.

Griffiths and Marjoram (1996) derived a recurrence equation similar to (25) allowing recombinations. A Markov chain Monte-Carlo approach similar to that by Griffiths and Tavaré for estimating $\theta$ (see the previous section) was developed for estimating $\theta$ and $C$. This approach appears promising but requires a substantial computer resource for even a modest value of $C$.

**6. Estimation of the age $T$ of the MRCA.** Estimation of $T$ is of particular interest in the study of human evolution, because the age of the most recent common ancestor (MRCA) of the modern humans is highly relevant to resolving the debate on the origin of modern humans.

The age $T$ of the MRCA of a sample is often referred to as the coalescent time of a sample and is given by

$$T = t_2 + \cdots + t_n \tag{47}$$

which can be seen from Fig. 1.

The distribution of $T$ without considering the information of polymorphism in a sample, therefore referred to as the *prior distribution*, can be found analytically, and has been derived by several authors. For example, Tajima (1990) found that the density function of $T$ is

$$\phi(T) = \sum_{k=2}^{n} (-1)^k (2k-1) \left( \prod_{i=1}^{k-1} \frac{n-i}{n+i} \right) k(k-1) e^{-k(k-1)T} , \tag{48}$$

which was given implicitly in Tavaré (1984) and was also derived by Fu (1996b).

Since the prior distribution of $T$ is the same for all the samples of the same size, and since we are interested in the value of $T$ for a particular sample or a population, we need to incorporate the information of polymorphism in a sample to obtain an estimate of $T$ for that sample. The first attempt of estimating $T$ using coalescent theory is due to Templeton (1993), who suggested the following estimator

$$\widehat{T} = \frac{\theta(1 + k_{max})}{2\mu(1+\theta)} , \tag{49}$$

where $k_{max}$ is the maximum number of nucleotide differences between two sequences in the sample, i.e., the number of nucleotide differences between the two most divergent sequences in the sample. Templeton's estimate is based on Tajima's (1983) result that in a sample of two sequences, the expected value of $T$ given the number ($\Pi$) of nucleotide differences between the two sequences is

$$E(T|\Pi) = \frac{\theta(1+\Pi)}{2\mu(1+\theta)} . \tag{50}$$

Since this expectation is valid only for two sequences (see, e.g., Fu and Li 1997), Templeton's approach is not a rigorous method. Indeed, Fu and Li (1997) (also see Tavaré et al. 1997) showed that Templeton's estimate is generally biased and the variance suggested by Templeton (1993) is also biased.

In general, estimates of $T$ can be derived from two probabilities. One is the probability $P(Data|T)$ (the likelihood function) of observing the data (the polymorphism ) given the value of $T$, and the other is the probability $P(T|Data)$ (the posterior probability) of a value of $T$ given the data. The value of $T$ that maximizes $P(Data|T)$ is the maximum likelihood estimate of $T$. Several estimators can be derived from $P(T|Data)$ and they are referred to as Bayesian estimators. We describe three such estimators below.

The mean estimator, $T_{mean}$, is defined as the mean value of $T$ given data. i.e.,

$$T_{mean} = E(T|Data) . \tag{51}$$

The mode estimator, $T_{mode}$, is the value of $T$ that maximizes the value of the posterior probability $P(T|Data)$. In addition to these two point estimates, we can define an interval estimate of $T$. The 95% (for example) interval estimate of $T$ is an interval within which the probability of $T$ falls is 95%. One such 95% interval estimate of $T$ is $(T_{2.5}, T_{97.5})$ where $T_{2.5}$ and $T_{97.5}$ are, respectively, the 2.5% and 97.5% percentile points of the distribution $P(T|Data)$ (see Fu and Li 1996 and Fu 1996b). A more conventional interval estimate is the so-called highest density set (O'Hagan 1994).

An important question is what type of estimator should one use ?. Figure 4 gives an example of the pattern of the three point estimators based on the number $K$ of segregating sites and shows that the maximum likelihood estimate gives a very small estimate of $T$ when $K$ is close to zero and a large estimate when $K$ is large. The maximum likelihood estimation is a well established statistical method and gives the best estimate in most situations. However, this method ignores the information prior to the sample collection, which is unjustified in the estimation of $T$ because a well-defined prior distribution of $T$ exists. In fact, a serious problem in the maximum likelihood approach arises when there is no variation in the sample. When the mutation rate is not zero (which can be verified using sequence comparison between different species), the maximum likelihood estimate of $T$

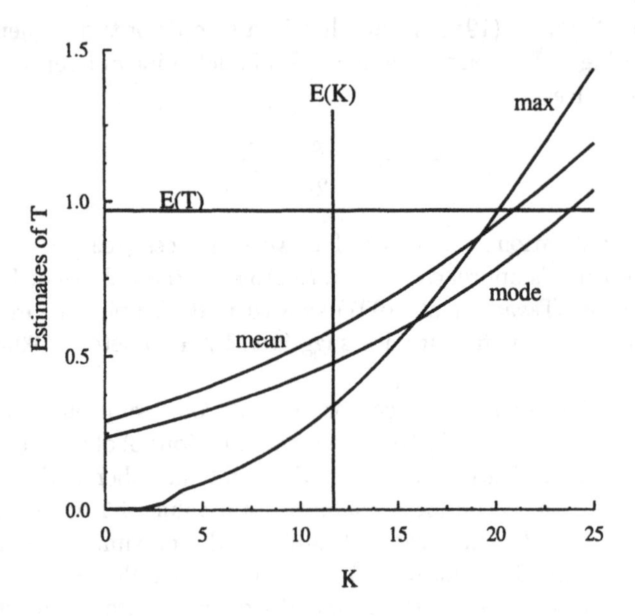


FIG. 4. *Estimate of $T$ (in unit of $4N$ generations) based on the number of segregating sites $K$ in the case of $n = 30$ and $\theta = 5$ (from Fu 1996b).*

is equal to 0 for a sample with no variation. Consider a non-recombining region, (for example, the mitochondrial DNA). There is unique value of $T$ for a sample regardless of the sequence length for the region studied. However, the shorter the sequences are, the larger the probability of observing no variation in the sample and the larger chance of having $T$ estimated to be 0 by maximum likelihood methods. This is obviously not a correct inference. ***When there is no such information from a sample, $T$ should be governed by its prior distribution.*** An example of a sample from an intron region of *ZFY* gene without polymorphic site was reported by Dorit et al. (1995) and how such a sample should be analyzed was discussed by Fu and Li (1996), Donnelly et al. (1996) and Weiss and von Haeseller (1996).

We now discuss methods for computing the posterior probability $P(T|data)$. We have used the term "data" loosely to denote the polymorphism information in a sample. The polymorphism in a sample can be represented by a number of ways, for example, the number $K$ of segregating sites, the maximum number $k_{max}$ of nucleotide differences between two sequences, the pattern of segregating sites, and the full information. We shall present the first and last cases.

Since

$$P(T|K) = \frac{P(K,T)}{P(K)}, \tag{52}$$

where $P(K)$ is the probability of $K$ segregating sites which can be computed from (24), we only need to derive the joint probability $P(K,T)$ of $K$ and $T$. Conditioning on the coalescent times $t_2,...,t_n$ the joint probability is given by

$$P(T,K|t_2,...,t_n) = \frac{e^{-\mu L}}{K!}(\mu L)^K \,, \tag{53}$$

where $L = 2t_2 + ... + nt_n$ is the total time length in the genealogy of a sample. The unconditional probability $P(T,K)$ can be found by integrating $t_2, t_3, ..., t_n$ with respect to their distributions, resulting in

$$p(K,T) = \frac{\theta^K n!(n-1)!}{K!} \int \cdots \int_{t_2+..+t_n=T} L^K \prod_k e^{-k(\theta+k-1)t_k} dt_n \cdots dt_2 \,. \tag{54}$$

Fu (1996b) showed that this probability is given by

$$p(K,T) = \frac{\theta^K n!(n-1)!}{K!} \sum_{k=2}^{n} \sum_{l=0}^{K} \alpha_{kl} \; k^l T^l e^{-k(\theta+k-1)T} \,, \tag{55}$$

where

$$\alpha_{kl} = \frac{K!}{l!}\beta_k(\theta)\gamma_{K-l,k} \,, \tag{56}$$

$$\beta_k(\theta) = \frac{(-1)^k(\theta+2k-1)}{(k-2)!(n-k)!\prod_{i=1}^{n-1}(\theta+k+i)} \,, \tag{57}$$

and

$$\gamma_{K-l,k} = \sum_{j_2+\cdots+j_n=K-l;\; j_k=0} \prod_m \frac{1}{(\theta+k+m-1)^{j_m}} \,. \tag{58}$$

In the case of no variation ($K = 0$), we have

$$p(0,T) = n!(n-1)! \sum_{k=2}^{n} \beta_k(\theta) e^{-k(\theta+k-1)T} \,. \tag{59}$$

Since Watterson (1975) showed that $p(K=0) = \prod_{k=1}^{n-1} \frac{k}{\theta+k}$, the posterior probability $p(T|0)$ is therefore

$$p(T|0) = n! \left[\prod_{k=1}^{n-1}(\theta+k)\right] \sum_{k=2}^{n} \beta_k(\theta) e^{-k(\theta+k-1)T} \tag{60}$$

which was derived explicitly by Fu and Li (1996) and implicitly by Donnelly et al. (1996). Fu (1996b) developed an iteration procedure for computing $P(K,T)$. An alternative procedure based on Monte-Carlo simulation was proposed by Tavaré et al. (1996).

Griffiths and Tavaré (1994a) extended their Markov chain Monte-Carlo method to estimate the joint probability of $T$ and sample configuration such as $(\boldsymbol{S}, \boldsymbol{n})$. The idea is similar to that for estimating $P(\boldsymbol{S}, \boldsymbol{n})$ but requires more computer resource. Once $P(T, (\boldsymbol{S}, \boldsymbol{n}))$ is estimated, the posterior probability of $T$ can be obtained by

$$P(T|(\boldsymbol{S}, \boldsymbol{n})) = \frac{P(T, (\boldsymbol{S}, \boldsymbol{n}))}{P(\boldsymbol{S}, \boldsymbol{n})} . \tag{61}$$

The accuracy of estimating $T$ using all available information in a sample should be higher in principle than those based on partial information. However, since evaluation of (61) involves approximations to the probabilities $P(T, (\boldsymbol{S}, \boldsymbol{n}))$ and $P(\boldsymbol{S}, \boldsymbol{n})$, part of the gain in using more information is lost. When the level of polymorphism is not very high, it may not be necessary to go beyond the estimate of $T$ based on $K$ alone, or $K$ plus the number of alleles (Griffiths and Tavaré 1996). It should be useful to identify the conditions at which more computer-intensive methods are worth pursuing.

One caveat in the estimation of $T$ is that information about mutation rate $\mu$ and the effective population size $N$ are required. $\mu$ is usually estimated from inter-specific sequence variation and $N$ is from independent estimate. In humans, chimpanzee sequences are commonly used to calibrate the mutation rate and $N$ is commonly taken as 10000, as estimated by Takahata (1993). Uncertainties in the values of $\mu$ and $N$ further complicate the inference on $T$. It is possible to incorporate the distribution of $\mu$ and $N$ formally into the estimation of $T$. For example, Donnelly et al. (1996) and Tavaré et al. (1996) considered a log-normal distribution for $\mu$. However, it is unclear at the moment what distributions are appropriate for $\mu$ and $N$.

**7. Estimation of ancestral population sizes.** We will discuss two methods, one developed by Takahata et al. (1995) and the other by Wakeley and Hey (1997).

Suppose species A and B diverged $t$ units of time ago and the effective population size at divergence was $N$. The coalescent time for two random orthologous sequences, one from each species, equals $t + s$, where $s$ is the coalescent time of two randomly selected sequences in a single population with effective population size equal to $N$. The probability that there are $k$ mutations separating two random sequences from two species is

$$P(k|s) = \frac{[2\mu(t+s)]^2}{k!} e^{-2\mu(t+s)} , \tag{62}$$

where $\mu$ is the mutation rate per sequence. Integrating $s$ over $(0, \infty)$ with respect to its exponential density results in the probability of $k$ as

$$P(k) = \left(\frac{1}{1+\theta}\right) \left(\frac{\theta}{1+\theta}\right)^k e^{-2\mu t} \sum_{i=0}^{k} \frac{1}{i!} \left(\frac{2\mu t(1+\theta)}{\theta}\right)^i , \tag{63}$$

where $\theta = 4N\mu$. If sequence pairs from $p$ independent orthologous regions are available, the joint probability is the product of $P(k)$'s over different regions. A maximum likelihood estimate can thus be obtained based on (63). Takahata et al. (1995) studied three pairs of species, human-chimpanzee (13 pairs of sequences), human-gorilla ( 7 pairs of sequences) and chimpanzee-gorilla (7 pairs of sequences). Assuming that the mutation rate per site per year is $10^{-9}$ and that one generation corresponds to 15 years, they estimated that $N = 83,000$ and $77,000$, respectively, for the ancestral population of humans and chimpanzees and the ancestral population of humans and gorillas. However, the estimate of $N$ using the chimpanzee and gorilla pair yields only $42,000$, although it is statistically not significantly different from $77,000$. This suggests that many independent pairs of sequences are required to obtain an accurate estimate.

The above method also yields estimates of the divergence time between species. Takahata et al. (1995) also developed a method for using information from three species . Yang (1997) showed, using simulations, that rate heterogeneity among different regions can significantly affect the estimate of the ancestral population size, although Takahata et al. (1995) was aware of this problem.

Wakeley and Hey (1997) considered the case in which two populations have diverged recently and samples of size larger than 2 from both populations are available. Let $\theta_1$, $\theta_2$, and $\theta_A$ be the $\theta$ values for populations 1, 2 and the ancestral population, respectively, and let $\tau = \mu t$, where $t$ is the divergence time and $\mu$ is the mutation rate per sequence per generation. Wakeley and Hey considered four types of segregating sites: $S_{X1}$, $S_{X2}$, $S_S$, and $S_f$. The first two are those specific to the sample from one population only. For example, $S_{X1}$ are segregating sites in the sample from population 1, but not in the sample from population 2. $S_S$ are those sites that are shared by both populations, and $S_f$ are those sites that are fixed between the two populations.

Wakeley and Hey (1997) developed a method for computing the expected values of $S_{X1}, S_{X2}, S_S$, and $S_f$. These expectations are complex functions of sample sizes, $\theta_1$, $\theta_2$, $\theta_A$, and $\tau$, but can be computed numerically. Substituting $S_{X1}, S_{X2}, S_S$ and $S_f$ for their respective values in these functions, one can solve a set of equations for values of the four parameters. The solutions are the estimates of the parameters. These estimators appear to be reasonably accurate when samples from multiple regions are available, but fail if samples from only one non-recombining region is available.

**8. Estimating the rate of selfing.** Until recently, coalescent approaches have been used mainly in animal population studies. Part of the reason is that many plant populations are partially selfing and existing methods do not take this into consideration. However, several recent developments (Slatkin 1991, Milligan 1995, Nordborg and Donnelly 1997, and Fu 1997a ) are providing more rigorous theory and methods for plant population studies. Although the development is still in its infancy, methods for estimating the selfing rate have received much attention, and we shall summarize them below. We will use Wright's (1969) definition of partial selfing, which assumes that each individual in a diploid population with selfing rate $s$ has probability $s$ of being the offspring of a self-fertilization and probability $1 - s$ of being the offspring of a random mating.

One method for estimating $s$ due to Milligan (1995) is as follows. Suppose a sample of $n$ sequences is taken from $k$ ($n/2 \le k$) individuals of a partially selfing population. Let $S_b$ and $S_w$ be, respectively, the mean nucleotide differences between two sequences from different individuals, and from the same individual. Extending Slatkin's (1991) results, Milligan (1996) showed that

$$E(S_b) = (1 - s)\theta \ , \tag{64}$$

$$E(S_w) = \frac{2 - s}{2}\theta \ , \tag{65}$$

which suggest the following simple estimates for $\theta$ and the selfing rate $s$:

$$\hat{\theta}_m = 2S_b - S_w \ , \tag{66}$$

$$\hat{s}_m = 2\frac{S_b - S_w}{2S_b - S_b} \ . \tag{67}$$

However, one problem with the above estimators is that they do not guarantee positive values while the true values of $\theta$ and $s$ are always non-negative. The proportion of samples resulting in negative $\hat{s}_m$ is quite high, rendering this estimate useless in many situations (Nordberg and Donnelly 1997, and Fu 1997a). For example, Fu (1997a) found from simulations that the percentage of samples resulting in negative $\hat{s}_m$ is 34% when $(n, k) = (20, 10)$ and $\theta = 5$, and 16% when $(n, k) = (80, 40)$. Even $\hat{\theta}_m$ can be negative, although the chance is much smaller. When this does occur, $\hat{s}_m$ becomes larger than 1, which is meaningless.

In a partially selfing population $S_b$ is expected to be larger than $S_w$. When the reverse is observed, resulting in either a negative value or a value larger than one for $\hat{s}_m$, it suggests that there is no evidence of partially selfing. This reasoning prompted Fu (1997a) to propose the following estimator

$$\hat{s}_f = \begin{cases} 0, & \text{if } S_b \le S_w \ , \\ 2\dfrac{S_b - S_w}{2S_b - S_b} \ , & \text{otherwise.} \end{cases} \tag{68}$$

Fu (1997a) also derived the following formula for the number $K$ of segregating sites from a sample of configuration $(n, k)$:

$$E(K) \approx \left[\frac{1}{2}a_n s + a_k(1 - s)\right]\theta \ , \tag{69}$$

where $a_n$ and $a_k$ are given by (14). An estimator of $\theta$ analogous to Watterson's for a random mating population is as follows:

$$\hat{\theta}_w = \frac{K}{a_n \hat{s}_f/2 + a_k(1-\hat{s}_f)} . \tag{70}$$

Fu (1997a) showed that both $\hat{s}_f$ and $\hat{\theta}_f$ are nearly unbiased, and that they both have considerably smaller variances than Milligan's estimates regardless of sample size.

Nordberg and Donnelly (1997) proposed a maximum likelihood method for estimating $s$ and $\theta$, which is based on the Monte-Carlo method by Griffith and Tavaré (1994c,1995). However, their approach is too time-consuming to be practical. For example, they reported that the computation for a sample of 20 individuals took days or even weeks on a workstation. Furthermore, they found that for a small sample size, their maximum likelihood estimates performed even worse than Milligan's estimates. These facts plus a rough comparison of their results with those in Fu (1997a) suggest that $\hat{s}_f$ and $\hat{\theta}_f$ are very efficient estimators. Nevertheless, further development of efficient inference methods is essential.

For the purpose of estimating selfing rate $s$, it is a better strategy to have a larger sample size than longer sequences because increasing sample size reduces the sampling variance more effectively. For a fixed total number of sequences, it is better to sample both alleles of every individual than to sample one allele of some individuals and both alleles of the remaining individuals (Fu 1997a).

**9. Testing the hypothesis of selective neutrality.** Whether a gene or a DNA region is free of selective constraints and evolves neutrally is always of great interest among evolutionists. There are two different versions of the neutrality hypothesis. One postulates that the majority of mutations that have contributed significantly to the genetic variation in natural populations are neutral or nearly neutral (Kimura 1968, 1983). This hypothesis is in general referred to as the *neutral mutation hypothesis* (see Li 1997). Another version assumes that the locus in question evolves according to the Wright-Fisher model and all mutations are selectively neutral. This later version is sometimes confused with the first one in the literature. To distinguish them, we shall refer to the latter as the *hypothesis of strict neutrality*. The availability of fast simulation algorithms from coalescent theory allows construction of statistical tests of the hypothesis of strict neutrality. We shall review statistical tests that utilize only the intra-specific polymorphism. A review of the tests using both intra- and inter-specific polymorphisms was given in Li (1997).

Tajima (1989) proposed to use the two different estimates $\hat{\theta}_T = \Pi$ and $\hat{\theta}_W = K/a_n$ of $\theta$ to detect selection. His test statistic is

$$T = \frac{\Pi - K/a_n}{\sqrt{Var(\Pi - K/a_n)}}, \tag{71}$$

where $a_n$ is given by (14). The rationale for this test is as follows. Since $K$ ignores the frequency of mutants, it is strongly affected by the existence of deleterious alleles, which are usually kept in low frequencies. In contrast, $\Pi$ is not much affected by the existence of deleterious alleles because it considers the frequency of mutants. Thus, if some of the sequences in the sample have selective effects, the estimate of $\theta$ based on $K$ will be different from that based on $\Pi$. Therefore, the difference $\Pi - K/a_n$ can be used to detect the presence of selection. The denominator is intended to normalize the test. Tajima suggested the use of a beta distribution as an approximation of the distribution of the test so that critical values of the test can be obtained, but Monte-Carlo simulation would be a more accurate method for obtaining the critical values of the test (Fu and Li 1993a).

One can construct a test of the same type as $T$ for any pair of estimates of $\theta$ as long as the variance of their difference can be calculated. However, for such a test to be useful, the two estimates must be sufficiently different when selection is present. Let $\eta_E$ and $\eta_I$ be the numbers of external and internal mutations, respectively ( note that $\eta_E = \xi_1$). Because the expectations of $\eta_E$ and $\eta_I/(a_n - 1)$ are both equal to $\theta$ under the neutral Wright-Fisher model and because they are likely to be different when selection is present, Fu and Li (1993a) proposed the following test:

$$D = \frac{\eta_I - (a_n - 1)\eta_E}{\sqrt{Var[\eta_I - (a_n - 1)\eta_E]}} . \tag{72}$$

When deleterious mutations are frequent and purifying selection is in action, most of the deleterious mutants will be eliminated from the population and those which are present at the time of sampling are most likely to have arisen recently and exist in low frequencies. Recent mutations are close to the tips (external nodes) in the genealogy and therefore are mostly included in the value of $\eta_E$. In contrast, mutations in the internal branches are most likely to be neutral and $\eta_I/(a_n - 1)$ is not strongly affected by the presence of selection. Therefore, these two estimates of $\theta$ should be useful for testing the presence of selection. Fu and Li (1993a) found that that $\eta_E$ and $\eta_I$ are weakly correlated, whereas $\Pi$ and $\hat{\theta}_W$ are strongly correlated. Tajima (1997) argued, however, that the magnitude of correlation is not a good indicator of the power of a test.

Because $\Pi$ is less affected than $\eta_E$ by the presence of selection, an obvious variant of test D is

$$F = \frac{\Pi - \eta_E}{\sqrt{Var[\Pi - \eta_E]}} . \tag{73}$$

The variances of $\eta_E - \eta_I/(a_n - 1)$ and $\Pi - \eta_E$ are given by Fu and Li (1993a). Note that since $\eta_E$ is much more strongly affected than $\eta$ by the presence of selection, we expect that $F$ is at least as powerful as Tajima's test $T$ for detecting the presence of selection.

When an outgroup sequence is not available, it is more convenient to consider the number, $\eta_s$, of singleton segregating sites instead of the number of external mutations. The expectation of $\eta_s$ was shown by Fu and Li (1993a) to be $[n/(n-1)]\theta$. The analogous tests to $D$ and $F$ using $\eta_s$ instead of $\eta_E$ are

$$D^* = \frac{\dfrac{n}{n-1}\eta - a_n\eta_s}{\sqrt{Var[\frac{n}{n-1}\eta - a_n\eta_s]}}\,, \tag{74}$$

$$F^* = \frac{\Pi - \frac{n-1}{n}\eta_s}{\sqrt{Var[\Pi - \frac{n-1}{n}\eta_s]}}\,. \tag{75}$$

Critical values of tests $D, F, D^*$ and $F^*$ have been obtained by Monte-Carlo simulations and are given in Fu and Li (1993a).

Since the variance and covariance of $\eta_i$ and $\xi_i$ are known, the mean and variance of any linear function of $\eta_i$ and $\xi_i$ can be computed and so can the covariance between any pair of linear functions of $\eta_i$'s or $\xi_i$'s. Therefore, many tests of the form

$$\frac{L_1 - L_2}{\sqrt{Var(L_1 - L_2)}} \tag{76}$$

can be constructed where $E(L_1) = E(L_2) = \theta$ under the neutral model. Tajima's test and Fu and Li's (1993a) tests are all this type of test. The difficulty is, however, to find the right pair of $L_1$ and $L_2$ such that the resulting test is powerful. Fu (1997b) discussed several other tests of this form.

Statistical tests can also be constructed using Ewens' sampling formula. Let $k$ be the number of haplotypes in a sample. Ewens (1972) and Karlin and McGregor (1972) showed that the probability of $k$ is given by

$$p(k) = \frac{|S_k|\theta^k}{S_n(\theta)}\,, \tag{77}$$

where $S_n(\theta) = \theta(\theta-1)\cdots(\theta-n+1)$ and $S_k$ is the Stirling number of the first kind, i.e., the coefficient of $\theta^k$ in $S_n$. Strobeck (1987) suggested to use the following statistic

$$St = \sum_{k\le k_0} \frac{|S_k|\Pi^k}{S_n(\Pi)} \tag{78}$$

for detecting population subdivision, where $k_0$ is the number of haplotypes in a sample and $\Pi$ is Tajima's estimate of $\theta$. It follows that $S$ is the probability of having $k_0$ or fewer alleles in a sample. The rationale is that when a population is subdivided, $\Pi$ will be inflated but $k_0$ is relatively unaffected. Therefore, using $\Pi$ as a substitute for $\theta$ will result in too few alleles in a sample, suggesting that the neutral model should be rejected.

Although Strobeck's test was designed for detecting population structure, it can be used to detect departures from the neutrality that result in a similar pattern of polymorphism as that of subdivided population, which is characterized by excess of old mutations or common alleles, or reduction of recent mutations or rare alleles. Several population genetics models, such as balancing selection and population shrinkage, have this characteristic. Fu (1996a) developed three new statistical tests for detecting such departure. Test $W$ is defined as

$$W = \sum_{k \le k_0} \frac{|S_k| \hat{\theta}_W^k}{S_n(\hat{\theta}_W)}, \tag{79}$$

where $\hat{\theta}_W$ is Watterson's estimate of $\theta$. The two other new tests $G_\xi$ and $G_\eta$ are based on mutations of various types. They are respectively

$$G_\xi = \frac{1}{n-1} \sum_{i=1}^{n-1} \frac{(\xi_i - \hat{\theta}_E/i)^2}{Var(\xi_i)}, \tag{80}$$

$$G_\eta = \frac{1}{\left[\frac{n}{2}\right]} \sum_{i=1}^{n-1} \frac{(\eta_i - \hat{\theta}_E/\alpha_i)^2}{Var(\eta_i)}, \tag{81}$$

where $[n/2]$ is the largest integer contained in $n/2$ and $Var(\xi_i)$ and $Var(\eta_i)$ are computed by formulas of Fu (1995) with $\theta$ replaced by Ewens' estimate of $\theta$. Fu (1996a) showed that these three tests are in general more powerful than Strobeck's test, Tajima's test, and Fu and Li's (1993a) tests for detecting departures that are characterized by excess of common alleles or reduction of rare alleles.

Population expansion, genetic hitchhiking (Maynard Smith and Haigh 1974) and background selection (Charlesworth et al. 1993) exhibit patterns of polymorphism that are the opposite of those of population subdivision and balancing selection. In other words, these three model often result in excess of rare alleles or mutations in a sample. Several simulation studies have been reported on the powers of various tests for detecting such departures from neutrality. Simonsen et al. (1996) found that Tajima's test is more powerful than Fu and Li's tests for detecting genetic hitchhiking, whereas Braveman et al. (1996) did not find significant difference in powers between these tests in most of the parameter combinations they examined. Charlesworth et al. (1996) found that Fu and Li's test $D$ is more powerful than Tajima's test in a limited simulation. Golding (1997) investigated the effect of directional selection on the lengths of internal and external branches of the sample genealogy and suggested that Fu and Li's (1993) tests may not have sufficient power to detect such selection. Fu (1997b) also conducted computer simulations to compare the powers of several existing and new statistical tests. Define

$$S = \sum_{k \geq k_0} \frac{|S_k|\Pi^k}{S_n(\Pi)} . \tag{82}$$

Fu (1997b) found that the new test

$$F_s = \ln\left(\frac{S}{1-S}\right) \tag{83}$$

is particularly powerful for detecting genetic hitchhiking and population expansion. He also confirmed that Tajima's test is indeed more powerful than Fu and Li's test in detecting genetic hitchhiking, but Fu and Li's tests $D$ and $F$ are considerably more powerful than any other existing tests for detecting background selection. The reason is that although background selection reduces the amount of polymorphism, the amount of reduction compared to the neutral model is similar for all types of mutations except for $\xi_1$ and $\eta_1$, that is, external mutations and singletons. Since Fu and Li's test $D$ (and $F$) compares $\theta$ values estimated from external (singleton) mutations and internal (non-singleton) mutations, the difference between the two estimates of $\theta$ is most striking under these two tests. Thus, they are more powerful than Tajima's test.

Since recombination reduces the variance of a linear function of $\xi_i$'s or $\eta_i$'s, tests of type (76) are conservative when there is recombination. On the other hand, recombination increases the number of alleles in a sample and thus reduces the power of Strobeck's test and the $W$ test. Since a larger number of alleles may lead to a significant result for test $F_s$, it should be cautious when there might be recombinations in a sample. Programs for performing the statistical tests described in this section can be obtained from Fu's webpage http://hgc.sph.uth.tmc.edu/fu.

**10. Prospects.** We presented an introduction to some mathematical aspects of the coalescent theory and reviewed some recent progress in both theory and applications. Our focus was on those that are helpful for understanding the mechanism of generating and maintaining molecular polymorphism in a population and for understanding the history of a population. The coalescent theory is widely regarded as the cornerstone for analyzing population samples and it is still progressing rapidly. Since there are huge arrays of questions of different nature and complexity about the evolution of a population or populations, which can be studied in the framework of coalescent, it is unlikely that one single inference approach is suitable for all purposes. Therefore, we expect to continue to see developments of various inference methods, as well as coalescent theory in models that have not been studied.

One area of the field that needs considerable improvement is user-friendly computer software for performing various analyses of population samples, although various programs are available from different authors. The current lack of user-friendly software is understandable because most of the sophisticated methods have only recently been developed.

One significant omission in our review is works on the variable number of tandem repeat (VNTR) type of data. Since human and many organisms have a huge number of polymorphic VNTR loci and many allele frequency data from these loci are now available, there is considerable interest in utilizing this type of data. A number of studies in the coalescent framework have been reported recently (e.g., Pritchard and Feldman 1996, Kimmel and Chakraborty 1996, Nielsen 1997, Slatkin and Ranala 1997, Fu and Chakraborty 1998). There is no doubt that much progress will be made in the near future.

**11. Acknowledgments.** This work was supported in part by NIH grants R01 GM55759(W.H.L), R01 GM30998 (W.H.L), R29 GM50428 (Y.X.F) and R01 HG01708 (Y.X.F and W.H.L), and NSF grant DEB-9707567 (Y.X.F).

## REFERENCES

[1] BRAVERMAN, J.M., R.R. HUDSON, C.H. KAPLAN, N.L. LANGLEY and W. STEPHAN (1995), *The hitchhiking effect on the site frequency spectrum of DNA polymorphisms,* Genetics **140**: 783–796.

[2] CHARLESWORTH, B., M.T. MORGAN and D. CHARLESWORTH (1993), *The effect of deleterious mutations on neutral molecular variation,* Genetics **134**: 1289–1303.

[3] CHARLESWORTH, D., B. CHARLESWOTH and M.T. MORGAN (1995), *The pattern of neutral molecular variation under the background selection model,* Genetics **141**: 1619–1632.

[4] CLEGG, M.T. (1997), *The Wilhelmine E. Key, 1994 invitational lecture, plant genetic diversity and the struggle to measure selection,* J. Hered. **88**: 1–7.

[5] COTTERMAN, C.W. (1940), *A calculus for statistico-genetics. Disertation,* Ohio State University, Columbus.

[6] DENG, W.H., and Y.X. FU (1996), *The effects of variable mutation rates across sites on the phylogenetic estimation of effective population size or mutation rate of DNA sequences,* Genetics **144**: 1271–1281.

[7] DONNELLY, P. (1986), *Partition structures, polya urns, the Ewens' sampling formula and the age of alleles,* Theor. Pop. Biol. **30**: 271–288.

[8] DONNELLY, P., S. TAVARÉ, D.J. BALDING and R.C. GRIFFITHS (1996), *Estimating the age of the common ancestor of men from the ZFY intron,* Science **272**: 1357–1359.

[9] DORIT, R.L., H. AKASHI and W. GILBERT (1995), *Absence of polymorphism at the ZFY locus on the human Y chromosome,* Science **268**: 1183–1185.

[10] ETHIER, S.N. and R.C. GRIFFITHS (1987), *The infinitely-many-sites model as a measure-valued diffusion,* Annals of probability **15**: 515–545.

[11] EWENS, W.J. (1972), *The sampling theory of selectively neutral alleles,* Theor. Pop. Biol. **3**: 87–112.

[12] EWENS, W.J. (1979), *Mathematical population genetics*, Berlin: Springer-Verlag.

[13] FELLER, W. (1968), *An introduction to probability: theory and applications*, volume 1, John Wiley & Sons, 3rd edition.

[14] FELSENSTEIN, J. (1992a), *Estimating effective population size from samples of sequences: inefficiency of pairwise and segregation sites as compared to phylogenetic estimates*, Genetical Research **56**: 139–147.

[15] FELSENSTEIN, J. (1992b), *Estimating effective population size from samples of sequences: a bootstrap monte carlo integration method*, Genetical Research **60**: 209–220.

[16] FU, Y.X. (1994a), *A phylogenetic estimator of effective population size or mutation rate*, Genetics **136**: 685–692.

[17] FU, Y.X. (1994b), *Estimating effective population size or mutation rate using the frequencies of mutations of various classes in a sample of DNA sequences*, Genetics **138**: 1375–1386.

[18] FU, Y.X. (1995), *Statistical properties of segregating sites*, Theor. Pop. Biol. **48**: 172–197.

[19] FU, Y.X. (1996a), *New statistical tests of neutrality for DNA samples from a population*, Genetics **143**: 557–570.

[20] FU, Y.X. (1996b), *Estimating the age of the common ancestor of a DNA sample using the number of segregating sites*, Genetics **144**: 829–838.

[21] FU, Y.X. (1997a), *Coalescent theory for a partially selfing population*, Genetics **146**: 1489–1499.

[22] FU, Y.X. (1997b), *Statistical tests of neutrality of mutations against population growth, hitchhiking and background selection*, Genetics **146**: 915–925.

[23] FU, Y.X. (1997c), *Probability of a segregating pattern in a sample of DNA sequences*, Theor. Pop. Biol. (in press).

[24] FU, Y.X. (1998), *Computing the probability of a segregating pattern in the infinite sites model*, Theor. Pop. Biol. (in preparation).

[25] FU, Y.X. and R. CHAKRABORTY (1998), *Simultaneous estimation of all the parameters of a stepwise mutation model*, Genetics (in press).

[26] FU, Y.X. and W.H. LI (1993a), *Statistical tests of neutrality of mutations*, Genetics **133**: 693–709.

[27] FU, Y.X. and W.H. LI (1993b), *Maximum likelihood estimation of population parameters*, Genetics **134**: 1261–1270.

[28] FU, Y.X. and W.H. LI (1996), *Estimating the age of the common ancestor of men from the ZFY intron*, Science **272**: 1356–1357.

[29] FU, Y.X. and W.H. LI, (1997), *Estimating the age of the common ancestor of a sample of DNA sequences*, Mol. Biol. Evol. **14**: 195–199.

[30] GOLDING, B. (1997), *The effect of purifying selection on genealogies*, In DONNELLY, P. and S. TAVARÉ, editors, *Progress in population genetics and human evolution*, 271–285, Springer.

[31] GRIFFITHS, R.C. (1989), *Genealogical tree probabilities in the infinitely-many-site model*, Journal of mathematical biology **27**: 667–680.

[32] GRIFFITHS, R.C. and P. MARJORAM (1996), *Ancestral inference from samples of dna sequences with recombination*, J. Comput. Biol. **3**: 479–502.

[33] GRIFFITHS, R.C. and S. TAVARÉ (1994a), *Sampling theory for neutral alleles in a varying environment*, Phil. Trans. R. Soc. Lond. B. **344**: 403–410.

[34] GRIFFITHS, R.C. and S. TAVARÉ (1994b), *Ancestral inference in population genetics*, Statistical Science **9**: 307–319.

[35] GRIFFITHS, R.C. and S. TAVARÉ (1994c), *Simulating probability distributions in the coalescent*, Theor. Pop. Biol. **46**: 131–159.

[36] GRIFFITHS, R.C. and S. TAVARÉ (1995), *Unrooted genealogical tree probabilities in the infinitely-many-sites model*, Math. Biosci. **127**: 77–98.

[37] GRIFFITHS, R.C. and S. TAVARÉ (1996), *Monte Carlo inference methods in population genetics*, Mathl. comput. modelling **23**: 141–158.

[38] HARDING, E.F. (1971), *The probabilities of rooted tree-shapes generated by random bifurcation*, Adv. Appl. Prob. **3**: 44–77.

[39] HASTINGS, W.K. (1970), *Monte Carlo sampling methods using Markov chain and their applications,* Biometrika **57**: 97–109.

[40] HEY, J. and J. WAKELEY (1997), *A coalescent estimator of the population recombination rate,* Genetics **145**: 833–846.

[41] HUDSON, R.R. (1991), *Gene genealogies and the coalescent process,* In Oxford Surveys in Evolutionary Biology, Ed. by D. Futuyma and J. Antonovics **7**: 1–44.

[42] HUDSON, R.R. (1993), *The how and why of generating gene genealogies,* In TAKAHATA, N. and A.G. CLARK, editors, *Mechanisms of molecular evolution,* 23–36, Sinaur Ass.

[43] HUDSON, R.R. and N.L. KAPLAN (1985), *Statistical properties of the number of recombination events in the history of a sample of DNA sequences,* Genetics **111**: 147–164.

[44] HUDSON, R.R. and N.L. KAPLAN (1994), *Gene trees with background selection,* In GOLDING, B., editor, *Non-neutral evolution: Theories and molecular data,* 140–153, Chapman and Hall, London.

[45] KARLIN, S. and J.L. MCGREGOR (1972), *Addendum to a paper of W. Ewens,* Theor. Pop. Biol. **5**: 95–105.

[46] KIMMEL, M. and R. CHAKRABORTY (1996), *Measure of variation at DNA repeat loci under a general stepwise mutation model,* Theor. Popul. Biol. **50**: 345–367.

[47] KIMURA, M. (1968), *Evolutionary rate at the molecular level,* Nature **217**: 624–626.

[48] KIMURA, M. (1983), *The neutral theory of molecular evolution,* Cambridge University Press, Cambridge.

[49] KINGMAN, J.F.C. (1982a), *On the genealogy of large populations,* J. Applied Probability **19A**: 27–43.

[50] KINGMAN, J.F.C. (1982b), *The coalescent,* Stochastic Processes and their applications **13**: 235–248.

[51] KRONE, S.M. and C. NEUHAUSER (1997), *Ancestral processes with selection,* Theor. Popul. Biol. **51**: 210–237.

[52] KUHNER, M.K., Y. YAMATO and J. FELSENSTEIN (1995), *Estimating effective population size and mutation rate from sequence data using Metropolis-Hastings sampling,* Genetics **140**: 1421–1430.

[53] LI, W.H. (1997), *Molecular Evolution,* Sinauer.

[54] LI, W.H. and Y.X. FU (1994), *Estimation of population parameters and detection of natural selection from DNA sequences,* In GOLDING, B., editor, *Non-neutral evolution: Theories and molecular data,* 112–125, Chapman and Hall, London.

[55] LUNDSTROM, R., S. TAVARÉ and R.H. WARD (1992), *Modeling the evolution of the human mitochondrial genome,* Math. Biosc. **112**: 319–335.

[56] MALÉCOT, G. (1941), *Étude mathématique des populations " mendéliennes",* Ann. Univ. Lyon Sci. Sec. A **4**: 45–60.

[57] METROPOLIS, N., A.W. ROSENBLUTH, M.N. ROSENBLUTH, A.H. TELLER and E. TELLER (1953), *Equations of state calculations by fast computing machines,* J. Chem. Phys. **21**: 1087–1092.

[58] MILLIGAN, B.G. (1996), *Estimating long-term mating systems using DNA sequence,* Genetics **142**: 619–627.

[59] NAGYLAKI, T. (1989), *Gustave Malécot and the transition from classical to modern population genetics,* Genetics **122**: 253–268.

[60] NEI, M. (1987), *Molecular Evolutionary Genetics,* Columbia University Press.

[61] NEUHAUSER, C. and S.M. KRONE (1997), *The genealogy of samples in models with selection,* Genetics **145**: 519–34.

[62] NIELSEN, R. (1997), *A likelihood approach to populations samples of microsatellite alleles,* Genetics **146**: 711–716.

[63] NORDBORG, M. and P. DONNELLY (1997) *The coalescent process with selfing,* Genetics **146**: 1185–1195.

[64] O'HAGAN, A. (1994), *Kendall's Advanced Theory of Statistics 2b: Bayesian Inference*, Edward Arnold: London.

[65] PRITCHARD, J.K. and M.W. FELDMAN (1996), *Statistics for microsatellite variation based on coalescence*, Theor. Popul. Biol. **50**: 325–44.

[66] SAWYER, S., D. DYKHUIZEN and D. HARTL (1987), *Confidence interval for the number of selectively neutral amino acid polymorphisms*, Proc. Nat. Acad. Sci. U.S.A. **84**: 6225–6228.

[67] SIMONSEN, K.L., G. CHURCHILL and C.F. AQUADRO (1995), *Properties of statistical tests of neutrality for DNA polymorphism data*, Genetics **141**: 413–429.

[68] SLATKIN, M. (1991), *Inbreeding coefficients and coalescent times*, Genet. Res. **58**: 167–175.

[69] SLATKIN, M. and B. RANNALA (1997), *Estimating the age of alleles by use of intra-allelic variability,*, Am. J. Hum. Genet. **60**: 447–58.

[70] STROBECK, C. (1983), *Estimation of the neutral mutation rate in a finite population from DNA sequence data*, Theor. Pop. Biol. **24**: 160–172.

[71] STUART, A. and J.K. ORD (1991), *Kendall's Advanced theory of statistics*, volume 2, Oxford University Press, 5th edition.

[72] TAJIMA, F. (1983), *Evolutionary relationship of DNA sequences in finite populations*, Genetics **105**: 437–460.

[73] TAJIMA, F. (1989), *Statistical method for testing the neutral mutation hypothesis by DNA popymorphism*, Genetics **123**: 585–595.

[74] TAJIMA, F. (1990), *Relationship between DNA polymorphism and fixation time*, Genetics **125**: 447–454.

[75] TAJIMA, F. (1993), *Measurement of DNA polymorphism*, In TAKAHATA, N. and A.G. CLARK, editors, *Mechanisms of molecular evolution*, 37–59, Sinaur Ass.

[76] TAJIMA, F. (1997), *Estimation of the amount of DNA polymorphism and statistical tets of the neutral mutation hypothesis based on DNA polymorphism*, In DONNELLY, P. and S. TAVARÉ, editors, *Progess in population genetics and human evolution*, 149–164, Springer.

[77] TAKAHATA, N. (1991), *A trend in population genetics theory*, In KIMURA, M. and N. TAKAHATA, editors, *New aspects of the Genetics of molecular evolution*, 27–47, Springer-Verlag.

[78] TAKAHATA, N. (1993), *Allelic genealogy and human evolution*, Mol. Biol. Evol. **10**: 2–22.

[79] TAKAHATA, N., Y. SATTA and J. KLEIN (1995), *Divergence time and population size in the lineage leading to modern humans*, Theor. Pop. Biol. **48**: 198–221.

[80] TAVARÉ, S. (1984), *Line of descent and genealogical process and their applications in population genetics models*, Theor. Pop. Biol. **26**: 119–164.

[81] TAVARÉ, S., D.J. BALDING, R.C. GRIFFITHS and P. DONNELLY (1997), *Inferring coalescence times from DNA sequence data*, Genetics **145**: 505–518.

[82] TEMPLETON, A.R. (1993), *The Eve hypotheses: A genetic critique and reanalysis*, American Anthropologist **95**: 51–72.

[83] WAKELEY, J. and J. HEY (1997), *Estimating ancestral population parameters*, Genetics **145**: 847–855.

[84] WATTERSON, G.A. (1975), *On the number of segregation sites*, Theor. Pop. Biol. **7**: 256–276.

[85] WEISS, G. and A. VON HAESELER (1996), *Estimating the age of the common ancestor of men from the ZFY intron*, Science **272**: 1359–1360.

[86] WRIGHT, S. (1969), *Evolution and the Genetics of Populations, Vol 2, The Theory of Gene Frequencies* The University of Chicago Press, Chicago.

[87] YANG, Z. (1997), *On the estimation of ancestral population sizes of modern humans*, Genet. Res. **69**: 111–116.

# PHYLOGENIES: AN OVERVIEW

SUSAN P. HOLMES*

**Abstract.** This is an overview that aims to help statisticians access interesting problems developing in the biological literature on estimating and evaluating phylogenetic trees.

**Key words.** Bootstrap, cladistics, DNA, molecular evolution, parsimony, phylogeny, systematics, tree.

**1. Introduction.** Representation of biological families by trees predates Darwin's theory of evolution, although the latter gave such representations a true explanatory justification. For biologists, at each branch of the tree are situated separation events that split orders or families or genera or species. For example, the figure shows a classification by Haeckel, 1870. Neighbors on the tree share the same ancestor. Characters that are de-

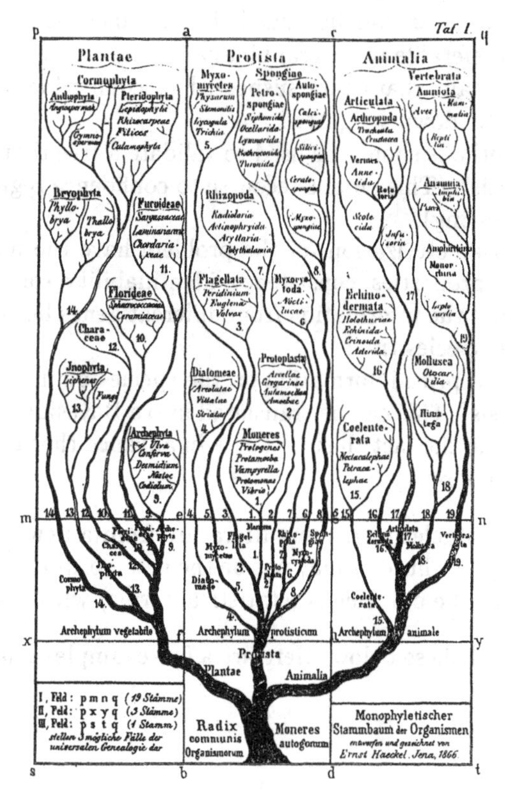


FIG. 1.

*Biometrics Unit, Cornell University, NY14853 Ithaca, `sph11@cornell.edu`

rived from this common ancestry are called homologous. Many geneticists doing population studies replace the term homology by identity by descent (IBD). The distinction between homology and similarity is a subtle one. In particular, sisters in the tree defined by a common ancestor are called clades or monophyletic groups, they have more than just *similarities* in common. Finding such groups is one of the goals of phylogenetic studies.

For over 200 years biologists have built trees to classify their species based on **morphological** [1] data. More recently the explosion of genetic data available through molecular biology has made tree-building even more popular. This presentation aims to interest statisticians in bringing their know-how to some of the open issues that currently fill the biological literature. The systematics literature is fraught with a great deal of polemics, much of which are statistical in nature.

Some questions that are raised include:

- Whether parametric methods using models should be preferred to nonparametric methods.
- How the data should be coded, for instance, is one categorical variable preferable to several binary ones?
- Should certain characters suspected of conflicting with the tree structure be down-weighted?
- Which methods should be used to validate a tree that results from the analysis? (This entails recourse to confidence regions and conditional testing).
- Which parameterizations of the problem have the most desirable statistical properties, consistency, identifiability, robustness ?
- How should the information from different genes be combined into an overall species tree?
- How should prior information on the species be incorporated into the analysis? This can be translated into a Bayesian dilemma; how can we code the fact that we know in advance that certain species are very different?

There are many obstacles to reading literature from a new field. Surprisingly, the most difficult hurdles may not be the new words encountered here but the 'faux amis'-the old friends (statistical terms) with new meanings.

I will document these below. Here are a few examples I had difficulties with:

[1] data about presence or absence of wings, sepals, hair, nodules,...

| Biological articles | | Standard Statistical Terminology |
|---|---|---|
| inferring phylogenies | - | estimating phylogenies |
| biased | - | systematically wrong |
| consistent | - | robust |
| consistency | - | existence of an iterative limit |
| statistical power | - | efficient |
| repeatability | - | |
| transition | - | |
| substitution model | - | transition matrix |
| independence | - | conditional independence |
| jackknife | - | cross-validation |
| statistical method | - | parametric method |
| likelihood | - | probability |
| statistician | - | philosopher |

Statisticians interested in more details about molecular evolution will find Li (1997) rewarding, it explains clearly many aspects of the problem. There is a collection of chapters on the subject in Hillis, Moritz and Mable (1996) which has the merit and handicap of attempting to be exhaustive. A review written for statisticians 15 years ago can be found in Felsenstein (1983). His programs are publicly available in `phylip` [24]. I will start my review as he did by defining phylogenies and the data used; then our paths separate. Section 3 presents a translation of the problem in statistical terms. Section 4 presents the three main families of tree-building methods: maximum likelihood, distance-based methods and maximum parsimony. Section 5 attempts to outline some of the sources of trouble in the procedures, and why 20 years after this field of research began, four specialized journals and hundreds of books later, no agreement has been reached, either on which method is better or how sure one is of the answer the methods provide.

Statisticians do have tools for comparing methods and Section 6 reviews some of the qualities of the various methods as measured with these statistical yardsticks. Section 7 presents methods for evaluating a tree, once it has been estimated. The question answered by the methods of this section are similar to those answered by the computation of a confidence region, unfortunately in a space with neither a natural distance nor a natural probability measure. The bootstrap is the most popular method among biologists for evaluating a tree, and we will try to underline some of its features and drawbacks in this context. Finally we will propose some more exploratory indices for evaluating how tree-like the data are to provide a scale of plausible error for the trees.

Finally the more practically minded may find the appendix a good starting point; it contains some exemplary runs of some freely available programs that can be easily down-loaded from the internet.

**2. What is a phylogeny?** ¿From a mathematical point of view a phylogeny is a rooted binary tree with labeled leaves.

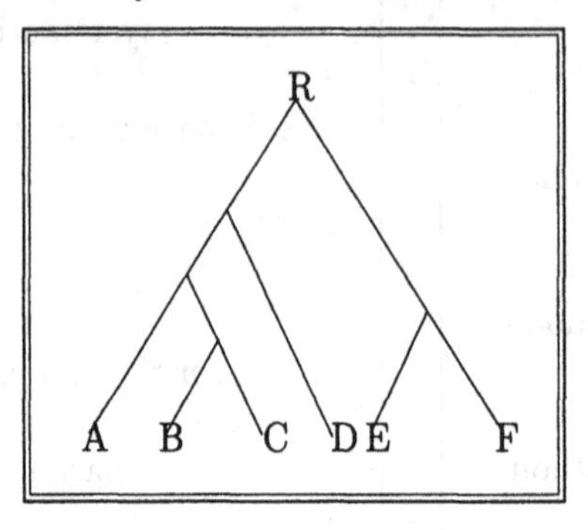


FIG. 2. *A rooted binary semi-labeled tree.*

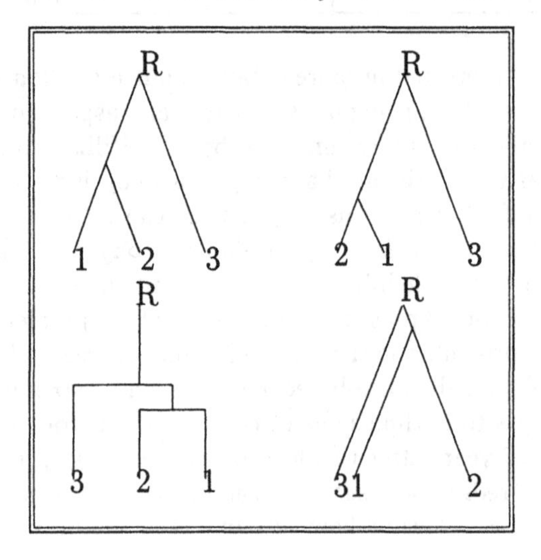


FIG. 3 *Four representations of a same tree topology.*

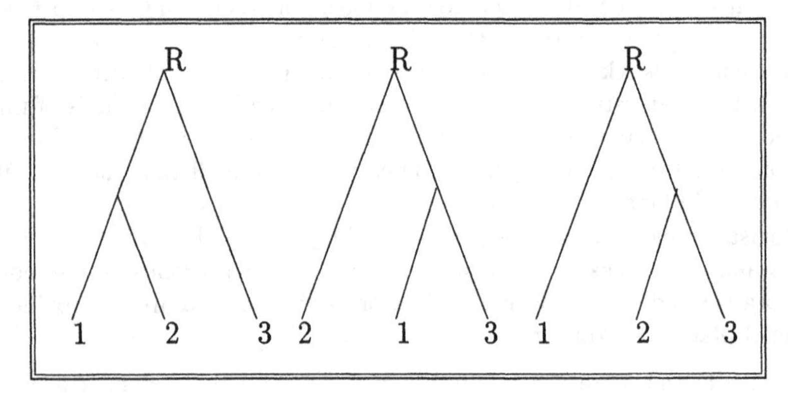


FIG. 4. *All possible distinct rooted binary topologies with 3 leaves.*

Unrooted trees are graphs in which all $N-2$ inner vertices (nodes) are of degree 3, and the $N$ outer vertices (leaves) of degree 1, are labeled.

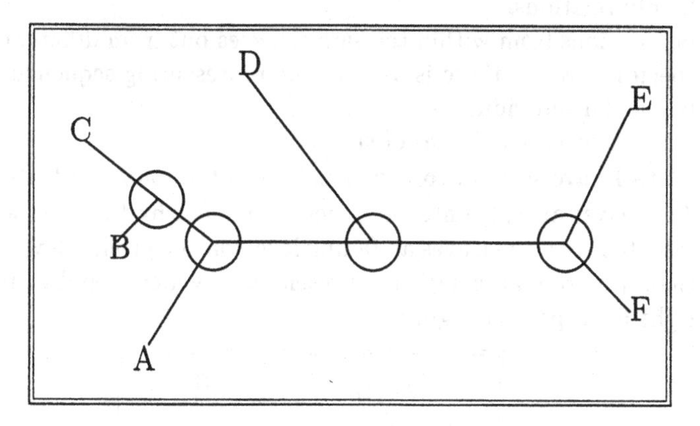


FIG. 5. *An unrooted tree.*

The number of unrooted semi-labeled trees with $N$ leaves is known since Schröder (1870) to be:

$$(2N-5)!! = (2N-5) \times (2N-7) \times (2N-9) \ldots \times 3 = \frac{(2(N-1))!}{2^{N-1}(N-1)!}$$

where $n!!$ is the double factorial where the difference between the successive factors is 2 instead of 1 in the classical factorial $n!$.

As there are $2N-3$ possible branches on which to place the root the number of rooted semi-labeled trees is: $(2N-3)!!$. For $N=10$ there are $2,027,025$ unrooted trees and $34,459,425$ rooted ones. These numbers grow rapidly. Using Stirling's formula, we have an asymptotic approximation

$$(2N-3)!! \sim (\frac{2}{e})^{(N-1)}(N-1)^{(N-1)}\sqrt{2}$$

This tells us that for $N = 20$, there are around $(2.10^{20})$ unrooted and $.8 \times 10^{22}$ rooted topologies from which to choose. Even if there is a lot of data, we can see that the choice is going to need some more outside information. We will develop this problem in section 6.

Many useful facts about such trees can be gleaned from books on graph theory and combinatorics: Stanley (1996) is particularly useful. It contains an elegant proof of Schröder's formula.

The leaves of these phylogenetic trees (called trees from here on) are called **Operational Taxonomic Units** or OTU's by the biologists and called simply units below. They can be:

- genes[2], for instance hemoglobins were some of the first to be sequenced and used for phylogenetic purposes.

[2]segment of DNA that codes for a polypeptide chain or specifies a functional RNA molecule, see Li (1997,page 9)

- individuals, represented by part of their genome. They are usually from within a population and can actually be connected by classical family relations.
- populations from within the same species but from different areas.
- species of which there is usually one representing sequence from a particular individual.
- families or larger classes of species.

The data I have seen up to now usually has the following features:

- The leaves are all contemporaries.[3] This is why the trees are represented with the leaves all falling level 'on the ground level' rather than a more mathematical representation which would inspire the right hand part of Figure 6.

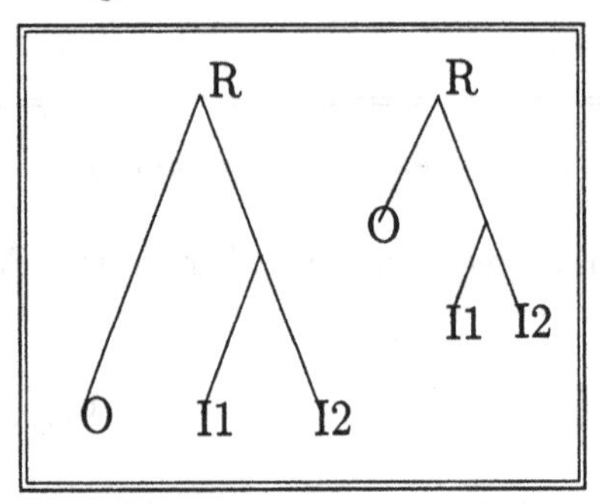


FIG. 6. *The left tree shows contemporary leaves.*

- Up to now sequencing has been slow and phylogenetic studies concentrated on relating species using one representative unit. With the rapid growth of Polymerase Chain Reaction (PCR) usage, data will be so abundant that it will be possible to study conjointly whole samples of sequences from different individuals of a same species, thus introducing interesting statistical information on variability.
- Usually it is the topology of the tree that is essential. The units (OTU's) are the only parts that are observed, the internal nodes have to be guessed at. (They are the ancestors, and they are the part of the tree that inspired the term *inferred.*) However, as we shall see later, the estimated tree is often augmented by branch-lengths. These present additional subtleties in the evaluation of the quality of the estimated tree.

**3. What are the data from which the trees are built?** We will suppose that we want to study $N$ units, this number of species or genes studied at the same time is usually between 10 and 50.

In all that follows the examples are of molecular data, either amino

[3]Excepting for paleontological data.

acids or nucleotides obtained sometimes from fragments, restriction sites, or whole DNA/RNA sequences[4].

The first part of the analysis consists of alignment of the sequences. There is a lack of coherence between the methods chosen to align the sequences and methods that are used after to build the tree. A method that would improve the coherence of the methodology would allow simultaneous alignment and tree-fitting. For recent work, see Schwikowski and Vingron (1996), and Wheeler (1994).

There are three main schools of tree-building methods:

- Maximum likelihood methods.
- Distance methods.
- Maximum parsimony methods.

For the time being I will consider that we are given sequences in which gaps have been inserted to enhance their **alignment** so that a typical sequence looks something like:

```
       21   383
VVi        M-SGTAGQVICCKAAVAWEAGKPVIEEVEVAPPQAMEVRLKILYTSLCH
Zma1       M--ATAGKVIKCKAAVAWEAGKPSIEEVEVAPPQAMEVRVKILFTSLCH
Zma2       M--ATAGKVIKCRAAVTWEAGKPSIEEVEVAPPQAMEVRIKILYTALCH
Hvu1       M--ATAGKVIKCKAAVAWEAGKPTMEEVEVAPPQAMEVRVKILFTSLCH
Hvu2       M--ATAGKVIKCKAAVAWEAGKPSMEEVEDAPPQAMEVRDKILYTALCH
Hvu3       M--ATAGKVIKCKAAVAWEAGKPSIEEVEVAPPQAMEVRVKILYTALCH
Tae        M--ATAGKVIECKAAVAWEAGKPSIEEVEVAPPHAMEVRVKILYTALCH
Osa1       M--ATAGKVIKCKAAVAWEAGKPSIEEVEVA--KEMEVRVKILFTSLCH
Osa2       M--AT-GKVIKCKAAVAWEAGEASIEEVEVAPPQRMEVRVKILYTALCH
Ath        M-S-TTGQIIRCKAAVAWEAGKPVIEEVEVAPPQKHEVRIKILFTSLCH
Psa        M-SNTVGQIIKCRAAVAWEAGKPVIEEVEVAPPQAGEVRLKILFTSLCH
Fan        M-SSTEGKVICCRAAVAWEAGKPVIEEVEVAPPHPNVVRVKILYTSLCH
Tre        M-SNTAGQVIKCRAAVAWEAGKPVIEEVEVAPPQAGEVRLKILFTSLCH
Stu        M-STTVGQVIRCKAAVAWEAGKPVMEEVDVAPPQKMEVRLKILYTSLCH
Pgl        M-A-TAGKVIKCKAAVAWEAGKPSIEEVEVAPPQAMEVRVKILYTSLCH
Phy        MSSNTAGQVIRCKAAVAWEAGKPVIEEVEVAPPQKMEVRLKILFTSLCH
Pde        M-SSTVGKVIRCKAAVAWEAAKPSIEEVEVAPPQANEVRLRILFTSLCH
Pta        MASSTAGQVIKCKAAVAWAAGEPKIEEVEVAPPQAMEVRVKIHYTALCH
Fra        M-SSTEGKVICCRAAVAWEAGKPVIEEVEVAPPQANVVRVKILYTSLCH
Mal        M-SNTAGQVIRCRAAVAWEAGKPVIEEVEVAPPQANEVRIKILFTSLCH
Lyc        M-STTVGQVIRCKAAVAWEAGKPVMEEVDVAPPQKMEVRLKILYTSLCH
```

This table is a subset of a larger data matrix that was downloaded from GENBANK. It was originally submitted by Yokoyama (1995) for all the sequences except `Vitis Vinifera` which comes from Sarni-Manchado,

[4] For detailed technical explanations see Li ,(1997) or Hillis et al, (1996)

Verriès and Tesnière (1997) and is from an *adh* gene.

The first two numbers indicate the dimensions of the data matrix $X$. The first number here is 21 because there are $N = 21$ species being studied, the second integer indicates that there are $k = 383$ characters. (they represent either one of the 20 amino-acids or an insertion '-').

The actual tree-building analysis will be run on different subsets of the data depending on which of the methods is used to build the tree.

1. For maximum likelihood, the complete matrix of sequences are used. Even columns with no difference at all between the units contain information on the relative frequency of various characters.
2. On the other hand, parsimony methods only use 'informative' columns, (for a complete definition see Li (1997), page 113). Informative sites are those that enable differentiation between possible trees, in particular either monotypical sites, or sites that have all the same value except for one unit are **not** informative and so are left out of the data set.
3. Distance methods have an in-between strategy. In a first step all the data are processed to estimate the relevant parameters for the distance formula, then the distances are computed between units. Only the distance matrix is used after that.

Other types of data can be used for tree building instead of DNA sequences. These can be either presence/absence of characters coded in binary and morphological characters coded as categorical data. Gene frequency data were used in the past, but recent molecular studies at a more precise scale seem to have replaced them.

At least one of the taxonomic units has a special function. For a statistician it would be seen as a simple outlier: the biologists voluntarily include what they call an **outgroup** to locate the root of the tree. The root is situated by creating an unrooted tree and the edge that joins the outgroup to the other species will be the support for the root. This is a clever use of prior information that simplifies the problem considerably, (by a factor of $(2N - 3)$). What is less obvious to the outsider is why, once the root's position is decided upon, the biologists keep the outgroup in the data set - it seems to distort the image of the closer group (called the **ingroup**), in fact outgroups also provide information on the root's characters, and so on the ancestral states of the character. This seems to be a security check, if in fact the outgroups become misplaced or lost in the tree, then there are signs of trouble. Many methods have trouble as soon as 2 very different outgroups are present (this is named the **long branch attraction problem**), just as in regression two opposite outliers can completely redefine the regression line.

In fact molecular data from one particular gene will only provide information about a certain 'gene tree' and not necessarily about the more general unit (such as the whole species). Combining information from all these different gene trees remains an interesting statistical open problem that could be addressed with conjoint methods such as those developed by Lavit, Escoufier, Sabatier and Traissac (1994) or by meta-analysis type methods. Some work has been started on the subject by Doyle (1992) and more recently Page and Charleston(1997).

**3.1. What are molecular-based phylogenies used for?** Lists of possible answers as extracted from Hillis et al. (1996), include gene evolution, population subdivision, analysis of mating systems and heterozygosity, paternity testing, as well as studies of individual relatedness, geographic variation, hybridization, species boundaries. Details of these can be found in the useful textbooks: Li (1997), Hillis et al. (1996).Comparative methods such as those explained in Harvey and Pagel (1993) also use phylogenies along with other, possibly quantitative, information.

Many modern genetic studies aimed at mapping diseases use the notion of **identity by descent**. This is the same as the concept of homology in the case of a study restricted to a small population of individuals for which gene trees are constructed. Thus information about relationships between far cousins can enhance understanding of homologies.

**4. Statistical translation of the problem.** All tree-building methods are based on the assumption that an evolutionary tree is a relevant representation of the data, an assumption that we will need to make more precise as we advance.

The first goal is **estimation**, producing a tree $\hat{\mathcal{T}}$ on the basis of a data matrix $X_{N\times k}$ that estimates an unknown true tree $\mathcal{T}$. This is strangely called an **inference** problem by biologists whereas the statisticians would call it an estimation problem.

The second goal is to provide a **confidence statement** to associate to the estimator $\hat{\mathcal{T}}$. Currently this is done most often by bootstrapping-type methods that we summarize in section 6. Although definitely related to branch lengths, this aspect of the tree receives less attention, since most goals of phylogeny seem to be more qualitative than quantitative.

The schools of tree-building methods: maximum likelihood methods, distance methods, and maximum parsimony methods can be compared using the statistical paradigm in a way which clarifies their similarities and differences. From a statistical viewpoint these methods can be understood as being ordered by the **dimension** of the underlying parameter space:

- Maximum likelihood uses a parametric model containing from 1 to 12 parameters for the substitution rates and usually $(N-2)$ parameters for the branching times. (See section 4 for a detailed description of the method.)
- Distance based methods use the same parametric model for the substitutions and deduce from these rates 'evolutionary' distances between units. The distance matrix is then analyzed by hierarchical clustering type methods such as neighbor-joining (single linkage clustering) or unweighted pair-group with arithmetic mean (aver-

age clustering). Distance-based methods can be seen as intermediary containing both parametric and nonparametric components.
- As we will see shortly, basic maximum parsimony methods are actually based on building a binary Steiner tree with regards to Hamming distance. They are nonparametric methods where the main assumptions are :
  - The existence of a true evolutionary tree.
  - The independence of characters (columns of the $X$ matrix).
  - Comparable substitution rates across characters.

Connections between the methods follows from recent work of Tuffley and Steel (1997) who show that when the number of parameters in the model is increased to incorporate different mutation rates along sites and different rates along branches the maximum likelihood method becomes equivalent to the maximum parsimony method.

Note that as the number of parameters becomes larger than the number of estimates that the data can usefully provide, the method passes the limit of a parametric model and becomes nonparametric (or infinite dimensional).

**5. The tree-building methods.** Here I will give a brief introduction to the three main families of tree-building techniques. Details may be found in Li (1997) for instance. Distance-based methods and maximum likelihood use a special model for describing the process by which changes between sequences occur. This is the substitution model that I will describe first.

**5.1. The substitution model.** To be more precise I will only show the case where the data are DNA nucleotides: **purines ('A', 'G')** and **pyrimidines ('T', 'C')**. There are many types of substitution models, the simplest model is called the Jukes-Cantor model and supposes that any change of the nucleotides occurs at the same rate, whether from one type to another, (**transversion**), for instance from purines to pyramidines within each type, (**transition**), for instance from purines to purines. The rate matrix $Q$ is of the form:

$$Q = \begin{array}{c|cccc} & A & T & C & G \\ A & -3\alpha & \alpha & \alpha & \alpha \\ T & \alpha & -3\alpha & \alpha & \alpha \\ C & \alpha & \alpha & -3\alpha & \alpha \\ G & \alpha & \alpha & \alpha & -3\alpha \end{array}$$

The 12 parameter model is of the form

$$Q = \begin{array}{c|cccc} & A & T & C & G \\ A & - & \alpha_{1,2} & \alpha_{1,3} & \alpha_{1,4} \\ T & \alpha_{2,1} & - & \alpha_{2,3} & \alpha_{2,4} \\ C & \alpha_{3,1} & \alpha_{3,2} & - & \alpha_{3,4} \\ G & \alpha_{4,1} & \alpha_{4,2} & \alpha_{4,3} & - \end{array}$$

The substitution matrix gives the probability of the change of a nucleotide during a time $t$ as:

$$P(t) = e^{Qt}$$

In the case of the amino acids we would have bigger matrices ($20 \times 20$ instead of $4 \times 4$), but most of the other computations carry through.

**5.2. Distance based methods.** These methods are variants of cluster analysis, probably more familiar to statisticians. The aim is to reconstruct the distances as computed between the two sequences of the two species $x$ and $y$ by distances along the edges of the tree forming a path between $x$ and $y$.

First a distance matrix is constructed between the $N$ units in some way. These distances $d_{xy}$ are supposed to estimate the unknown 'true evolutionary' distances between $x$ and $y$ as they would be measured along the unknown true tree $\mathcal{T}$.

For the Jukes-Cantor model which assumes equal rates of substitution between all base pairs provides the estimate of distances between sequences $x$ and $y$ as:

$$d_{xy} = -\frac{3}{4}\log(1 - \frac{4}{3}(1 - (\frac{\#AA}{k} + \frac{\#CC}{k} + \frac{\#GG}{k} + \frac{\#TT}{k})))$$

where $k$ denotes the number of characters (columns) in the data matrix, and $\#AA$ denotes the number of times there is an $A$ in $x$ matched with an $A$ in $y$.

Once the distances are decided upon, the parametric model is left behind and a clustering technique such as hierarchical clustering with average groups is used to find the tree from the distances.

It seems that this method has declined somewhat in popularity over recent years among biologists. It was the method that made the trees the easiest to compute, but improved facilities have made maximum likelihood and maximum parsimony more tractable. Those who don't believe in the parametric substitution models don't use it because of the assumptions underlying the distance computations and those who don't trust heuristic tree-building algorithms don't use it because of the tree-building phase. Historically it was the first method available on the computer, and people still use it for reasons of computational ease.

**Remarks:**

If we knew the true evolutionary distances between species, we could build an additive tree that reproduced the distances along the tree in a unique way. The existence of an additive tree reproducing the distances faithfully is not always ensured, a sufficient condition for this to be possible is called the **four point condition**:

$$d_{AB} + d_{CD} \leq max(d_{AC} + d_{BD}, d_{AD} + d_{BC}), \text{for all quadruples } (A, B, C, D)$$

This means that one of the two sums is minimum and the other two are equal. Notice that this is not the same as the ultrametric property which says that for any three points: A, B, C:

$$d_{AC} \leq max(d_{AB}, d_{BC})$$

If the distances obey the ultrametric property the distances can be fit to a binary tree with leaves equally distant from the root. Unfortunately distances computed from real data never obey this property. We will give details later in section 5, but additivity is destroyed by:

- Homoplasy (reversal, parallelism and convergence) which is caused by superimposed changes.
- An uneven distribution of change rates.
- Measurement error.
- **Paralogous** sequences[5].

Some distances are obtained directly by hybridization techniques. We will not include these here. We concentrate on distances that are computed from substitution models such as Jukes and Cantor's one-parameter model, Kimura's two-parameter model, or even the complex 12-parameter model for the substitution matrices. These models provide estimates of differences between sequences computed from the frequencies of various changes in the sequences.

**5.3. Parsimony method.** The foundations of this method have been long discussed, as always with heuristic nonparametric procedures. A detailed account can be found in Farris (1983), his justification for parsimony is that this method "minimizes requirements of ad hoc hypotheses of homoplasy[6]". This is easier to understand for a statistician if the analogy is made between homoplasies and residuals, these are the part of the data that the tree does not explain, minimizing homoplasies is an approach akin to minimizing residuals in regression for instance.

Roughly this method can be seen as based on the assumption that "evolution is parsimonious" which means that there should be no more evolutionary steps than necessary. Thus the best trees are the ones that minimize the number of changes between ancestors and descendants. We will see that under the assumption of independence of each of the characters, this has a clear combinatorial translation.

**5.3.1. The parsimony tree as a combinatorial problem.** For the time being, we will only consider the construction of unrooted parsimony trees. As we saw in the section on data, the rooting of the tree is done before the construction of the unrooted tree.

[5] Consequences of lineages being created separately after a gene duplication.

[6] These are the transformations caused by reversal, parallelism and convergence that will be explained in section 5.

Recall that the Hamming distance between two units is the number of changes needed to bring one to the other. This assumes that all changes in a categorical character are counted as one step.

$$d_H(AACTGGG, AACTGGC) = d_H(AACTGGG, AACTGGA) = 1$$

Here, given $N$ points in a metric space, the Steiner problem is that of finding the shortest tree connecting the $N$ points where one is allowed to add extra vertices. Thus, with 4 points arranged at the vertices of a unit square, one would add a fifth point in the center to form the Steiner tree.

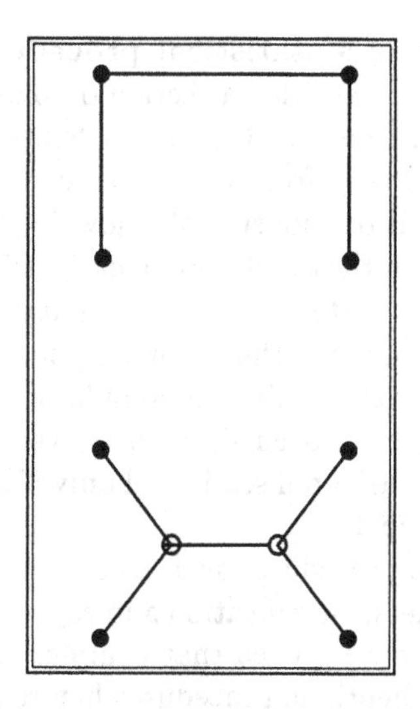

FIG. 7. *The minimum spanning tree and the Steiner tree of the 4 vertices of a rectangle.*

Although statisticians are not familiar with minimal Steiner trees, they may have encountered minimal spanning trees as used by Friedman and Rafsky (1985). The relation between the two is well explained in Gardner's wonderful chapter on Steiner trees (Chapter 22, Gardner (1997)). He explains how minimal spanning trees are good "starting points" since in the plane for instance they can only be 13% longer than Steiner trees.

As a combinatorial problem, the maximum parsimony tree is the problem of finding the Steiner points or Steiner tree for Hamming distance between the units, under the constraint that the tree be binary. The problem of finding a minimal Steiner tree is that of finding the Steiner points (representing ancestors) that minimize the complete length of the tree. Steiner points are points that are added to a graph so that its minimal spanning tree becomes shorter. The minimal Steiner tree problem is NP-hard, meaning that no algorithm is known that will compute an optimal tree in polynomial time in the number of species $N$.

Much work has been done to implement good heuristic algorithms for finding approximately optimum trees. Swofford's `PAUP`, Felsenstein's

`Phylip`, and Goloboff's `NONA` all contain clever use of branch and bound techniques and branch swapping to find acceptable answers. No explicit analyses of the complexity of the algorithms involved have been published, but recent empirical tests show enormous progress in terms of CPU time, even for large (N=500) problems (Goloboff, personal communication).

Theoretical computer scientists on the other hand have produced many papers on methods to solve the problem, with detailed complexity analysis, but no code is available as yet. See in particular recent work such as Erdös, Steel, Szèkely, Warnow(1996,1997) and Rice, Steel, Warnow and Yooseph (1997).

**5.3.2. Parsimony as a statistical procedure.** Felsenstein (1983) lists parsimony in a section entitled a section on parsimony as "non-statistical approaches". Farris says (1983) says the "statistical approach to phylogenetic inference was wrong from the start, for it rests on the idea that to study phylogeny at all one must first know *in great detail* how evolution has proceeded". Both these authors identify statistics with parametric modeling. This is unfortunate as it has led many clever cladists to stop reading the statistical literature, thus depriving them of many useful tools. Parsimony methods are well within the boundaries of non-parametric statistical procedures that have been developed over the last twenty years. Methods are no longer considered statistical only if they are justified by an underlying stochastic model.

Many data-analytic procedures such as correspondence analysis, projection pursuit, neural nets, classification and regression trees (CART) and minimal spanning trees have proved that complex situations can be satisfactorily understood by heuristic procedures before any theoretical framework supposing a probabilistic model justifies their properties (Diaconis and Efron (1984)).

On the other hand it can be an interesting challenge for theoretical statisticians to do for parsimony what Rubin and Anderson (1956) did for factor analysis, that is find a model for which the heuristic method was providing the correct estimate, as also was the case for partial likelihood. However I doubt from a practical point of view that this would be of any interest to those who use parsimony as their standard tree-building technique.

**5.4. Maximum likelihood trees.** For a statistician this is the easiest of the methods to understand. A parametric model $(\theta, T)$ is postulated, $\theta$ is a $\eta$-dimensional vector that we explain below and $\mathcal{T}$ is the tree's topology. Under this model the likelihood for each possible tree $\mathcal{T}$ is separately computed for each character, the independence of characters then allows the total likelihood of the tree for all data to be computed by taking the product.

The first part of the vector of parameters $\theta$ comes from the substitution model as explained in section 4.1. The number of other parameters that have to be specified depends on the complexity of the model. If a molecular clock [7] is postulated, speciation times $\{t_1, t_2, ...t_{N-2}\}$ (splitting events) are the other parameters. Otherwise both the branch lengths $\{v_1, v_2, ...v_{N-2}\}$ and the different rates along those branches have to be parametrized.

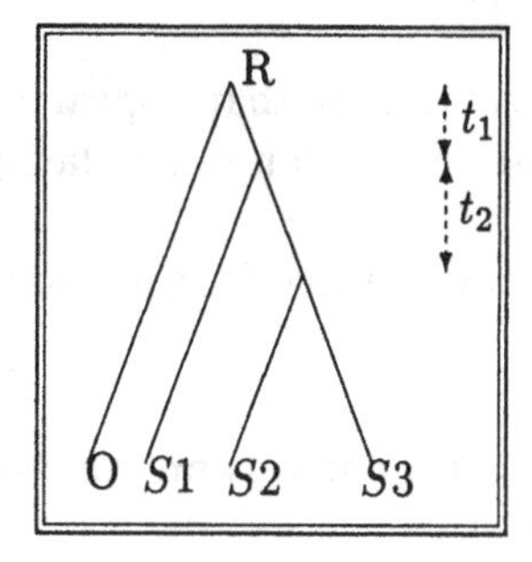


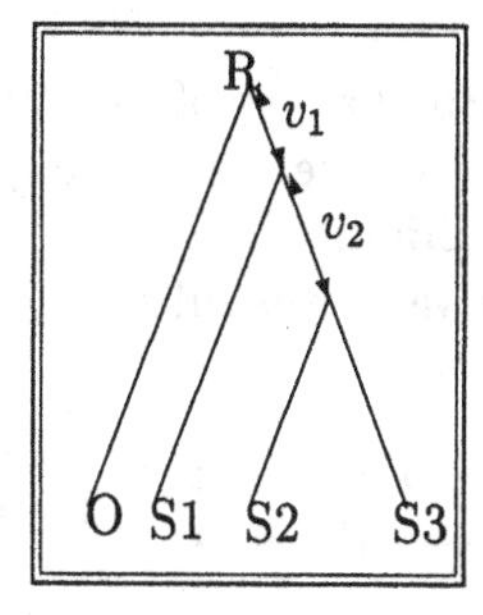


FIG. 8. *Two parametrizations of the tree.*

The substitution parameters are estimated from the data. A complete model including distributions of separation events is postulated and the likelihood can be computed for each possible tree by computing the likelihood of the tree given each site $X_{.j}$:

$$f(X_{.j}|\theta_1, \theta_2, \ldots, \theta_\eta, \mathcal{T}).$$

This actually requires computing the likelihood of all the subtrees, so the method is recursive.

$$\mathcal{L}(\theta_1, \theta_2, \ldots, \theta_\eta | X_{.1}, X_{.2}, \ldots, X_{.k}, \mathcal{T}) = \prod_{j=1}^{k} f(X_{.j}|\theta, \mathcal{T})$$

As the assumptions are essential, I present them here:

1. Each site in the sequence evolves independently.
2. Different lineages evolve independently.
3. Each site undergoes substitution at an expected rate which is chosen from a series of rates with a given distribution.

Fancier versions of the procedure enable different sites to have different evolution rates.

Many biologists won't use maximum likelihood because of the computational expense, each tree's likelihood computation is $NP$ hard. This is a surprising exception to the usual rule that parametric methods are advantageous by their lesser computational needs. Others don't use the MLE because there seems to be little evidence that the assumptions are actually realistic in real biological applications.

**6. Where does the trouble come from?** Here are a few details about the hurdles the tree-making algorithms have to deal with.

[7] branch lengths in evolutionary change depend linearly on time

**6.1. Homoplasy.** A character change may become invisible through time, because there has been a **reversal** or **back-substitution** for instance:

$$A \longrightarrow G \longrightarrow A.$$

There are also changes of exactly the same type that appear in different parts (clades) of the tree, giving a false impression of similarity. This is called **parallelism.**

Another variant is substitutions that occur in different clades but have the same results:

$$\left.\begin{array}{r} A \longrightarrow G \longrightarrow A \\ A \longrightarrow C \longrightarrow T \longrightarrow A \end{array}\right\} \text{ these are called \textbf{convergent} substitutions.}$$

The effect on the resulting measurements of differences between units are the same: there is an error; units appear to be more similar than they would be if the complete history were known. Collectively these are called **homoplasy**. There are very clearly documented examples of these in Li (1997), pages 69-70.

Parametric models that take homoplasy into account are the motivation for the 'modified evolutionary distance' computations. Whether they include 1 or 12 parameters they try to retrieve some of the variability lost through homoplasy. Some authors feel that this possibility of error-correction in parametric methods is so essential that it justifies using such models even when they have not been proved to fit the actual phenomenon.

Parsimony methods are sometimes limited to shorter stretches of time to limit the homoplasy; 'long branches' are undesirable in parsimony methods.

**6.2. Non-optimality of the solutions.** As we saw in section 4, both maximum likelihood and parsimony provide only locally optimum trees, whatever their criteria, because the problems are computationally intractable. The clustering methods used on distance matrices are also only heuristic algorithms, not necessarily providing the global optimum.

So even when the data are perfectly dependable, errors may persist because only a local optimum was obtained, or there may be several optima. Some authors repeat the analysis of the data, interchanging the order of their species, this makes the algorithm choose a different starting point, thus often resulting in a different solution. This appears as option `jumble` in `Phylip` for instance (see the examples in the appendix).

**6.3. Many possible trees, little data.** When we have boiled down the patterns of different nucleotides, there are often less than 100 of them left, of which usually 80% or more are weak signals because they are singletons. So we have about 100 numbers, the frequencies of each of the patterns, from which to decide about $10^{20}$ possible trees, a difficult task.

In more detail, although there may be $k = 2000$ characters available, most of these are usually uninformative sites. Even parametric methods that use them (parsimony doesn't) boil them down to 4-5 numbers: the number of columns of each type. For instance, in the `Vitis Vinifera` example of section 1, there were 383 columns of amino-acids of which 187 were all monotypical columns. When one takes out what biologists call **singletons**, (columns with all the same character but for one species) there were only 140 columns of data left. Now if the data are patterned, (the character which appears first is called '1', the next different '2', the third '3' and so on[8]), it can be summarized as a few frequencies, most equal to $\frac{1}{k}$.

**7. Evaluating the methods.** Some statistical yardsticks such as consistency, efficiency, identifiability, robustness, computational speed, discriminating ability, or versatility may help to compare the methods in an abstract way.

**7.1. Accuracy.** A first suggestion that comes to mind is "*Which method gives the true tree when we know the answer ?*" Unfortunately, there are few data sets where the truth is known. An example is the tiny organism called bacterophage T7 of which a small phylogeny was generated in laboratory conditions (Hillis, Bull, White, Badgett and Molineux 1992). The programs in the appendix show some of the results obtained on this data for which parsimony seemed to work well, it gave an accurate prediction of the tree. But a sample of size one is no evidence, and as usual the statistician begs the biologists: *Bring us more data!*.

**7.2. Consistency.** There have been studies of consistency of the estimator $\hat{\mathcal{T}}$ in the classical statistical sense: when the number of characters increases to infinity do the trees provided by the estimators converge to the true tree? Under their own particular assumptions, all methods are consistent. However this is insufficient unless these conditions can be checked. Chang (1996), shows that maximum likelihood is inconsistent when the homogeneity assumption of identical distribution of substitution rates across characters is violated. Parsimony is inconsistent when some branch lengths are long enough to make 'hidden changes' or homoplasies likely.

In fact, the justification of putting this into a classical statistical framework is tricky, because what is being said is that we should consider that characters can be independently sampled from some distribution. Then, as the number of characters increases, we want the estimator to converge to the true tree. However, such increase in the observations is impossible, the genome is finite, and as we sample more and more characters they are less and less independent. (see Sanderson 1995)

Consistency is a quality that should not be considered fundamental. We never have infinite amounts of data, especially as compared to the number of choices that have to be made, on the other hand, it would be

[8]For an example see in the appendix.

most useful to know how long a sequence is necessary to attain a sufficient level of precision in distinguishing between possible trees, this has been named **statistical power** in part of the biological literature. In fact a more precise statistical term would be **efficiency**.

**7.3. Efficiency.** Historically one could reason backwards and see why biologists have called this **power**, but as no specific testing framework is set up before the analysis, this term seems abusive here. In classical statistical terminology, **efficiency** measures how quickly a method converges to the correct solution as the data size increases, this would be a better term here.

Much theoretical work remains to be done here. For maximum likelihood, classical estimates of efficiency are available. No such information is available for nonparametric estimation methods.

**7.4. Robustness.** Robustness measures the stability of the method when the data do not fulfill the necessary assumptions. Simulations can be used to test robustness with regards to specific departures from the assumptions. There have been some of these done by biologists. No theory is available, in particular the notion of influence function needs distances to be defined in both tree-space and data-space. Neither have been studied in this context.

**7.5. Identifiability.** Making the maximum likelihood model more flexible to encompass more biological realism is blocked by the problem of non-identifiability. When both branch lengths and substitution rates are free to vary, the model ceases to be identifiable. This is studied in Chang (1995).Too many parameters and too little data are the plight of phylogeny. This will only get worse as one starts to study real biological data with all the dependency between characters included.

**8. Evaluating phylogenies.** Various questions that biologists need to answer after building a tree $\hat{\mathcal{T}}$ from their data are:

1. How sure am I that this clade exists?
2. Can I be more confident in clade $A$ than in clade $B$?
3. If the data were slightly wrong(a bad alignment, a poor reading of the characters), how far off would my tree be?
4. How much support does this clade have from the data?

These questions do not necessarily have to do with a precise stochastic setting as many authors have pointed out. For interesting reflexions on statistics in a non-stochastic setting see Freedman and Lane (1983). For a discussion of the foundation of the use of the bootstrap from the biological point of view see Sanderson (1995).

**8.1. The bootstrap.** To clarify some of the messy issues here I will try to develop a somewhat geometric analysis of the problem, this is more fully developed in Efron, Halloran and Holmes (1996).

Suppose that the number of characters is fixed at $k$ and the number of units or species is $N$. The data are $k$ characters from an alphabet of length $A$ (maybe $A = 5, \{A, G, C, T, -\}$ or $A = 21$ amino acids and a '-'). The set of all possible columns of $N$ species from an alphabet of size $A$ is $S = A^N$. Under the assumption that the columns of $X$ are exchangeable, a data matrix $X$ can be associated with a unique $\hat{\pi}$ , the vector of relative frequencies of each type of possible column. This is not an economic way of coding the data. The vector is of length $S$ and is extremely sparse, but it is conceptually useful here. The tree-estimation process associates to this data vector $\hat{\pi}$ an estimated tree $\hat{\mathcal{T}} = \mathcal{E}(\hat{\pi})$. Of interest are properties of the estimated trees $\hat{\mathcal{T}}^* = \mathcal{E}(\hat{\pi}^*)$ for neighboring $\hat{\pi}^*$.

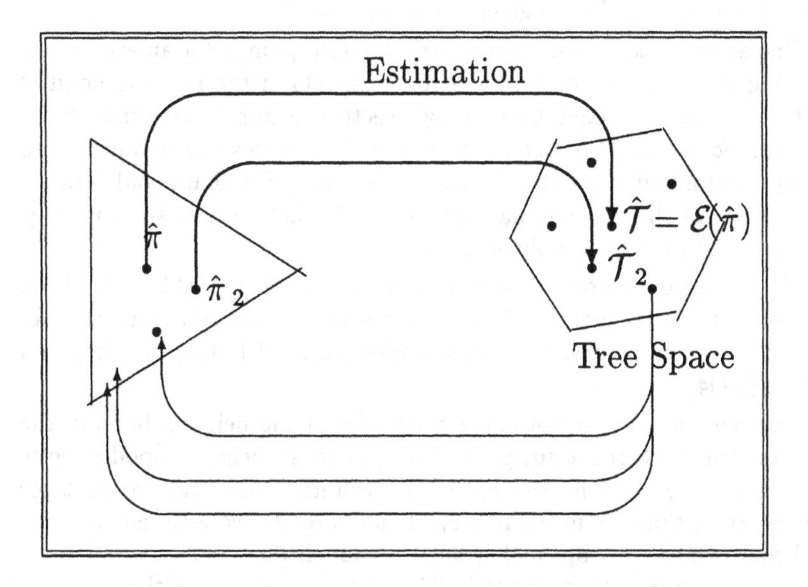


FIG. 9. *Estimation is a function from the data to tree space.*

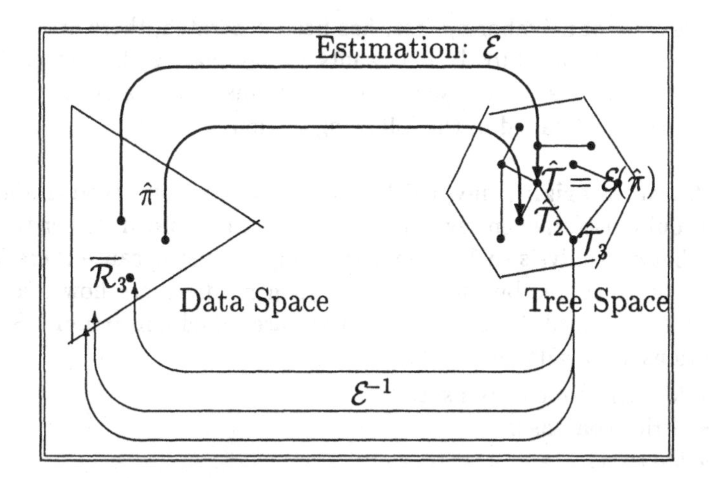


FIG. 10. *Several different data sets could give the same estimated tree.*

Nonparametric bootstrapping is a way of creating a neighborhood of close, plausible frequencies $\hat{\pi}^*$ by redistributing $k$ columns among all the observed columns. Looking at the associated trees provides 'neighboring trees'. Properties of the corresponding neighborhood of $\hat{\mathcal{T}}$ are supposed to represent properties of the neighborhood of the true tree $\mathcal{T}$.

We are interested in trees that are obtained as images of possible frequency vectors $\hat{\pi}^*$. The idea is to find out if for some $\hat{\pi}^*$ "near" $\hat{\pi}$, the tree $\hat{\mathcal{T}}^*$ is different from $\hat{\mathcal{T}}$.

For instance, in the second figure above, region $\mathcal{R}_3$ is the set of all frequencies that would have given the same tree $\hat{\mathcal{T}}_3$.

The above describes nonparametric bootstrapping. Parametric bootstrapping is also a way of studying some aspects of the neighborhood of $\hat{\mathcal{T}}$. This is done by generating new $\pi^*$ vectors as simulated through the relevant stochastic model taking the tree and the necessary parameters to be those estimated from the data set. **Seq-gen**, Rambaut and Grassly (1997),is one of the software packages available that enables such a study. (See appendix for an example of its use).

Simulating data from a given tree and stochastic model has also been much used to experiment with the nonparametric bootstrap in the absence of necessary theory. (See for instance Chernoff (1997) or Berry and Gascuel (1994)).

Another method for obtaining properties of the neighborhood of the estimated tree is **Bremer support**. This provides a neighborhood directly in tree space by relaxing the optimality criteria somewhat, so that for instance, trees that are up to 10 steps longer are also considered. Or, one could continue to relax optimality until a clade disappears. This was what Bremer (1988)originally suggested. This gives a measure of the diameter of a neighborhood around $\hat{\mathcal{T}}$ defined by contour lines of the function that is optimized.

Once a set of neighboring trees has been generated, there are different ways of using them. Mostly one wants to summarize the properties of these neighborhoods in tree space, again an unsolved statistics problem. One approach is detailed in the following section.

**8.2. Summarizing several trees.** The **consensus tree** is a notion which is quite useful when several trees have been obtained, either through a perturbation analysis such as bootstrapping, or just because there is not a unique optimal tree but several. One then needs to see how much the various trees concord. Two trees that agree are called **congruent**. Several propositions are available:

- Majority rule consensus.
- Strict consensus.
- Quartets.
- Compatible components.

Phylip offers either of the first two. The first creates a tree where the clades are those that have a majority of trees in their favor. In the second, strict consensus, only clades that have unanimous support are shown, others appear as 'unresolved'. An example output from Phylip can be found in the appendix.

Current preoccupations of biologists seem concentrated on how to split up the NP-hard problem of finding the optimal tree and recombine various partial solutions. One of the solutions proposed is to divide up the data at random in a cross-validation type procedure, and re-unite the trees with consensus methods. (This is called the parjack by Farris, et al. (1996)) Another is to use all the quartets and recombine them. This is called the **quartet puzzling** method and there is available software called puzzle, (Strimmer and von Haeseler, (1996)) which provides ways of doing this for the maximum likelihood criterion based trees. Certainly, more study is needed on combining trees.

Biologists exhibit bootstrap results by drawing a consensus tree. The number of times a given monophyletic group or clade appears on the tree is divided by the total number of trees simulated. This number is written along the branch of the consensus tree as an indication of how 'sure' one could be of the clade. In an abuse of nomenclature, it is called **bootstrap support**.

Interpretation of such a usage is particularly difficult for anyone who has had enough training in probability to use probability trees where the numbers along the branches are conditional probabilities.

This method is so popular with certain schools of biologists that any paper exhibiting a tree without "bootstrap support" numbers on the branches is rejected.

**8.2.1. Of the use of p-values?** The bootstrap support is often assimilated to a p-value, the technical discussion of such an interpretation has already been given elsewhere (Efron, Halloran and Holmes, 1996). Although one can ponder whether several p-values associated with a same tree with the same data set one shouldn't worry about the multiple testing aspect, I will only raise a philosophical issue here. Don Ylvisaker pointed out during the workshop that it is becoming customary in court cases to ask statisticians to stand up in court and state their 'p-values' as evidence. This seems to have replaced the notion of an expert. For recognizing fingerprints, the expert says whether the fingerprints were beyond the shadow of a doubt those of a certain person. Although tempting as it may be to quantify the 'beyond the shadow of a doubt' as a number (with eventually several decimal places of precision....) these $p$ values are in fact meaningless. We know that only approximate answers are possible.

**8.2.2. Why can the bootstrap run into trouble?** The first time I saw the use of the bootstrap in this context, it seemed that the role of variables and observations had been reversed as compared to the traditional setup. This may create confusion for statisticians accustomed to data matrices with few columns (variables) and many observations (rows).

Here are some other possible sources of error in bootstrapping:

1. Discreteness of the underlying statistic: the tree is a discrete statistic, for which no applicable theory exists for the use of the bootstrap with reasonable amounts of data. Large deviations as developed by Newton (1996) are unfortunately not applicable. Zharkih and Li (1995) defined the "partial bootstrap" that statisticians will recognize as a $m$-out-of-$n$ bootstrap that attempts to fix the problems that the bootstrap encounters when the estimated tree $\hat{\mathcal{T}}$ is close to several possible neighboring trees.
2. The statistic $\hat{\mathcal{T}}$ is based on a maximum. It is well-documented that bootstrapping doesn't work for maximums of random variables (Bickel and Freedman, 1982).
3. Overparametrization of the model compared to quantity of data available. In multivariate regression for example, the bootstrap fails completely when the number of variables, and so the number of parameters, becomes of the same order as the number of observations. Work by Freedman and Peters (1984)documents this carefully. Another well explained example can be found in the use of the bootstrap to estimate bias in classification and regression trees (CART), see Breiman and Stone (1984) who explain the magical $1 - \frac{1}{e}$ factor also rediscovered by Harshman (1994).
4. Non-independence of observations: The closer we want the model to adhere to real nucleotide data, the more one sees that the characters are not independent. The codons (triplets of DNA) have to be dependent as there are only certain ones that are possible, those that code for certain amino-acids. There is also well documented **secondary structure** [9] across the sequences which is also evidence against independence. The columns could be considered conditionally independent given the tree, but I have not found any literature explaining this different assumption. It seems that the dependence structure is precisely what one is trying to find in the tree structure.
   Statisticians will recognize here a wonderful field of application for methods of inference more precisely tailored for dependent data, that is, block-bootstrapping, Markov Chains, etc..
5. Non identity of the distribution at different states. Any graphical display such as can be seen in the appendix shows that there are regions where there are many changes and other more stable regions. This spatial dependency should be integrated into the bootstrap.

**8.3. Probability distribution on trees.** A statistician considering the inferential part of the analysis of trees would characterize how close we

[9]The sequences fold and parts react together to perform certain functions.

believe the estimate to be to the true tree by using sampling theory. This builds on a probability distribution on the space of all trees.

The difficult aspect of this problem is that there are exponentially many possible *trees* that the parameter can take on. The classical non-parametric approach to this would be to put a multinomial probability model on the whole set of trees. This would have dimension $d \sim N^N$, $N$ being the number of species. It could be possible to use a different parametric approach than the substitution model, using prior knowledge on the species' relations or the tree's form. The use of the outgroup strategy explained above is a special case of this. For instance, one could use functions such as the depth of the tree, the number of two-leaved clades, the balance of the tree, etc, and create exponential families through these. The more the parameters, the closer we can come to nonparametric models while keeping a hold on the overall structure of the tree.

Seen this way we can understand that as the maximum likelihood method puts a low dimensional surface through this high dimensional space, its chances of finding a tree 'near' the true tree may be quite low.

The notion of a tree that is *near* the true tree has not been discussed here, nor very much in any of the biological literature. Waterman and Smith define the NNI metric which is at the basis of the 'elementary steps' that Pearl, Doss, Li (1997)use in their Gibbs sampler for generating posterior distributions on tree space. The 'branch swapping' methods used both in PAUP (Swofford 1998) and NONA (Goloboff 1994) also use these elementary steps to explore parts of tree-space searching for optimal trees. We refer the interested reader to Diaconis and Holmes (1998) for reviews of possible distances on trees and random walks on the space of trees that have various desirable properties.

Defining a graph of trees that are nearest neighbors in some sense[10] can be useful. Distances between trees can be defined as the number of edges separting the trees in this graph. Random walk on trees can be seen as random walk on this graph.

Another idea that is possible when a distance between trees has been defined is to look for trees that are suboptimal, but close to **interpretable** trees.

The Bayesian approach advocated and carried through by Doss, Pearl and Li (1996) and Mau, Newton and Larget (1999) defines parametric priors on the space of trees, and then computes the posterior distribution on the same subset of the set of all trees. These enable precise confidence statements in a Bayesian sense.

**9. Exploratory indices.** For a statistician starting an analysis of aligned molecular data on $N$ species, a first question might be about the relevance of building a tree, or how far the data lie from a 'reasonable tree'.

[10]they may differ by the transposition of leaves, or by a pruning/reconnection move

That is basically how tree-like are the data? Of course given complete freedom, we can always build a tree that obeys certain rules and connects the species.

Each of the different tree-building contexts, parametric, semi parametric and nonparametric can be submitted to such an evaluation in a coherent way. We follow previous work from Sattvah and Tversky (1977) who study trees as compared to planar multidimensional scaling in the reconstruction of distances for the use of hierarchical clustering for psychological data who suggest the following indices of treelike-ness. For $d(i,j)$ the distance as measured by the distance matrix between units $i$ and $j$ and $d_{\hat{\mathcal{T}}}(i,j)$ the distance as measured along the tree.

$$STRESS = \sum \frac{|d(i,j) - d_{\hat{\mathcal{T}}}(i,j)|}{\sum d(i,j)}.$$

$STRESS$ measures how well the tree reconstructs all the distances.

Measures of how tree-like the data are include the consistency index, a measure used by parsimony-tree builders. It is defined as the ratio of the minimum number of steps a tree with $k$ characters and $N$ species would need, divided by the actual number of steps needed. Thus this index is always between 0 and 1.

Measures of tree-likeness provide size estimates for what could be considered a reasonable neighborhood within which to search for an interpretable tree. Again, there is much to do here.

**10. Conclusions.** There are suggestions that statistical theory can make to help biologists, here are some examples:

- An evaluation of some of the error could be made by finding how *far* the data are from being tree-like, this could indicate what *size* the neighborhood of possible trees should be. Here I have used notions of distance both in the data space and the tree-space without defining them, this should be a first step for theory (see Diaconis and Holmes, 1998).
- Verification of assumptions and quantification of notions of robustness should go hand in hand.
- There should be a coherence of methods for each part of the analysis. If there *is* an underlying model used in the alignment procedure, then the same model should be employed throughout the tree-building process and its validation. For instance a 2 parameter mutation model used to align sequences implies that a distance based method using Kimura's distance is appropriate. This should be followed by parametric bootstrapping using this same parametric model and software such as **Seq-gen** (Rambaut and Grassly, 1997) [11]. This is not circularity, it is coherence.

[11] Using as arguments the estimates of transition/transversion ratio and nucleotide frequencies provided by the data

On the other hand, if the method is nonparametric, alignment will probably be better done either by hand in an exploratory fashion or at least without a parametric model, validation methods can include nonparametric bootstrapping, although if independence between the branches is not assumed, there is no meaning to creating bootstrap numbers along the branches representing averages. It would seem more correct to give the more frequently obtained trees, with their probabilities.

- Monte-Carlo Markov chains for generating sampling distributions on tree space seems like an interesting one (Mau, Newton & Larget, (1999))
- This is a high dimensional problem, *curse of dimensionality* tells us there is NO reason to melt it down to just a planar representation with numbers along the branches without more ado.....

On the other hand there is much to be learnt from the clever algorithms that are being developed by cladists to attack this complex problem, for instance the successive weighting algorithm (Farris 1969) could be transposed into a statistical framework for regression. The procedure reweights the characters after the first tree is found, downweighting those that are discordant with it, and then repeating this until an optimal tree is found. This is like an iterated reweighted least-squares method. Goloboff (1997) has proposed a less computer intensive version of this that creates the weights once only, and assigns an overall cost to each tree taking the weights into account. This is like a downweighting least-squares method.

Those who have run simulation studies on constructing trees have noticed that the bootstrap combined with consensus methods has a propensity to correct bias. Thus Berry and Gascuel (1997) and Erdös, Steel, Szèkely and Warnow (1997) have rediscovered this property of the bootstrap's, already documented in the general case (see Efron and Tibshirani, 1993). Thus combining many bootstrap data sets and then taking a consensus tree has been shown empirically to produce a better estimate than just a one-pass parsimony optimization. However no theory has yet been developed in this case to explain and quantify the improvement.

Serious statistical considerations have also led to many other rediscoveries. For example, Zharkikh and Li (1995) rediscovered the merits of the $m$-out-of-$n$ bootstrap in the multiple decision context that occurs when there are more than two trees that are plausible given the data.

Certainly I feel that the two fields of evolutionary biology and statistics would gain in more interdisciplinary work.

**11. Acknowledgements.** Many thanks to those who have wasted their precious time reading and discussing this work; Herman Chernoff, Jerry Davis, Jeff Doyle, Persi Diaconis, Brad Efron, David Freedman, Wen-Hsiung Li and Kevin Nixon. Thanks to Brigitte Charnomordic who wrote the interactive alignment program, and to Tandy Warnow, Lászlo Székely and Joe Chang for sending me copies of their work. And thanks to Betz Halloran for inviting me to this IMA workshop which has been a wonderful opportunity to talk with colleagues.

Note added in proof: I also thank Joe Felsenstein who took the trouble of sending me many pages of comments that could unfortunately not be accomodated for because of time constraints.

## APPENDIX

**Examples of data set.** I have used two data sets for my examples, the T7 data experimentally generated phylogeny, Hillis et al. (1992) for which the parsimony program will be seen to produce the correct answer. Here is the part of the data set (in `phylip` form) composed of the informative sites:

```
     9  21
R          C C G C C G G C C G G C C A G C G G G G T
J          C C C C G T A C C G G T C A A C G G G G T
K          T C C C G C A C C G A T C A A T G G G G G
L          T C C C G C A C C G A T C A A T G G G G G
M          C T C C G T A C C G G T C A A C G G G G T
N          C C T T A C G T T A G C T G G C A A A A T
O          C T C C G C G C T G G C C G G C A G A A T
P          C C C C A C G C T G G C C G G C A G A A T
Q          C C T T A C G T T A G C T G G C A A A A T
```

If the data set is put into a file called infile it will automatically be processed by any `phylip` program that is called. Otherwise if there is no current infile, `phylip` will ask for a file name, then there is a dialogue menu that allows the user to specify all the options.

**Parsimony tree.** This is part the output from the **phylip** command **dnapars**:

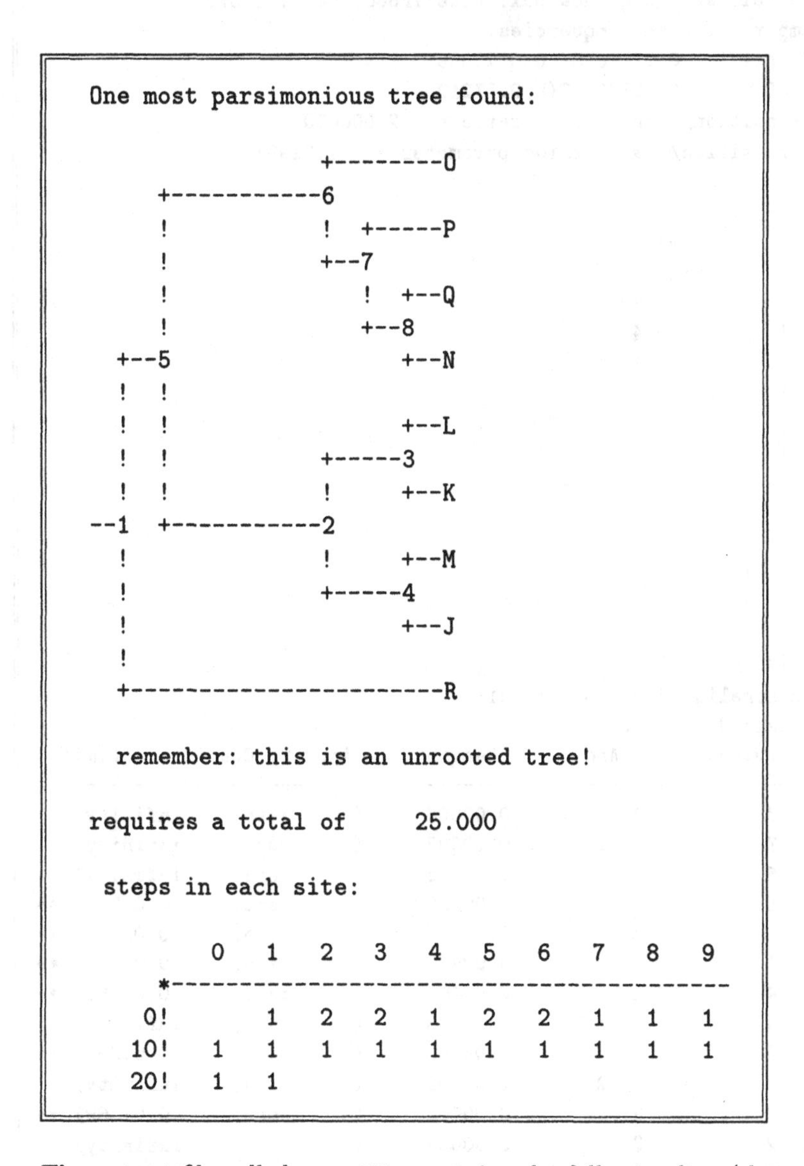

```
One most parsimonious tree found:

                 +--------O
     +-----------6
     !           ! +-----P
     !           +--7
     !              ! +--Q
     !              +--8
  +--5                 +--N
  !  !
  !  !                 +--L
  !  !           +-----3
  !  !           !     +--K
--1  +-----------2
  !              !     +--M
  !              +-----4
  !                    +--J
  !
  +-----------------------R

  remember: this is an unrooted tree!

requires a total of     25.000

 steps in each site:

         0   1   2   3   4   5   6   7   8   9
     *-----------------------------------------
    0!       1   2   2   1   2   2   1   1   1
   10!   1   1   1   1   1   1   1   1   1   1
   20!   1   1
```

The output file called **treefile** contains the following line (the tree in parentheses format):

```
(((O,(P,(Q,N))),((L,K),(M,J))),R);
```

**Maximum likelihood trees:** Output from phylip program dnaml:

```
Nucleic acid sequence Max. Likelihood, vers. 3.572c
Empirical Base Frequencies:
   A       0.27778   G 0.22685
   C       0.22325  T(U)0.27212
Transition/transversion ratio =   2.000000
(Transition/transversion parameter =   1.519971)
  +J
  !
  !       +R
  !    +--1
  !    !  !  +N
  !    !  +--4
  !    !     !  +O
  ! +--5     +--3
  ! !  !        !  +P
  ! !  !        +--2
--7--6 !           +Q
  ! !  !
  ! !  +L
  ! !
  ! +M
  !
  +K
Ln Likelihood =  -344.10331
Examined   95 trees
 Between         And         Length      Approx. Confidence Limits
 -------         ---         ------      ------- ---------- ------
   7          J            0.00006     (      zero,     infinity)
   7             6         0.00003     (      zero,     infinity)
   6             5         0.00006     (      zero,     infinity)
   5             1         0.00936     (      zero,      0.02236) **
   1          R            0.00466     (      zero,      0.01384) **
   1             4         0.00469     (      zero,      0.01389) **
   4          N            0.00462     (      zero,      0.01369) **
   4             3         0.00003     (      zero,     infinity)
   3          O            0.00462     (      zero,      0.01369) **
   3             2         0.00003     (      zero,     infinity)
   2          P            0.00462     (      zero,      0.01369) **
   2          Q            0.00003     (      zero,     infinity)
   5          L            0.00006     (      zero,     infinity)
   6          M            0.00003     (      zero,     infinity)
   7          K            0.00003     (      zero,     infinity)
   *  = significantly positive, P < 0.05
  ** = significantly positive, P < 0.01
```

**How tree-like were the data?** Here is the distance as computed by Jukes-Cantor distance between the bacteriophage species:

```
R   0   4   8   8   5  11   6   6  11
J   4   0   3   3   0  12   6   6  12
K   8   3   0   0   4  15  10  10  15
L   8   3   0   0   4  15  10  10  15
M   5   0   4   4   0  13   6   7  13
N  11  12  15  15  13   0   5   3   0
O   6   6  10  10   6   5   0   1   5
P   6   6  10  10   7   3   1   0   3
Q  11  12  15  15  13   0   5   3   0
```

as compared to the distances along the branches of the 'best' distance based tree:

```
     [,1] [,2] [,3] [,4] [,5] [,6] [,7] [,8] [,9]
[1,]    0    7   10   10    8   15    9    9   15
[2,]    7    0    5    5    1   16   10   10   16
[3,]   10    5    0    0    5   19   13   13   19
[4,]   10    5    0    0    5   19   13   13   19
[5,]    8    1    5    5    0   16   11   11   16
[6,]   15   16   19   19   16    0    8    6    0
[7,]    9   10   13   13   11    8    0    2    8
[8,]    9   10   13   13   11    6    2    0    6
[9,]   15   16   19   19   16    0    8    6    0
```

For which the stress was:

```
> sqrt( sum((d7a-d72f)^2)/sum(d72f^2))
[1] 0.1205607
```

**Parametric bootstrap generation of sequences.** Suppose we had the `treefile` from a previous phylip output, the generation of sequences is done using `Seq-gen` (Rambaut and Grassly, 1997) by :

```
seq-gen -mHKY -t3.0 -127 -n100 < treefile > example.T7
```

For which the output looks like:

```
Sequence Generator - seq-gen, Version 1.04
(c) Copyright, 1996 Andrew Rambaut and Nick Grassly
Department of Zoology, University of Oxford
South Parks Road, Oxford OX1 3PS, U.K.
Simulating 11 taxa, 27 bases
  for 1 tree(s) with 100 dataset(s) per tree
Branch lengths assumed to be number of substitutions
per site
Rate homogeneity of sites.
Model=HKY
  transition/transversion ratio = 3 (kappa=6)
  frequencies = A:0.25 C:0.25 G:0.25 T:0.25
0%|____________________|100%
  [....................]
Time taken: 0.12 seconds
```

The data file example.T7 generated looks like this:

```
 11 27
Pfa4      CCGACCTCCAAGATTCGCTATGACAAT
Pvi10     CCGACCTCCAAGATTCGCTATGACAAT
Pcy9      CCGACCTCCAAGATTCGCTATGACAAT
Pkn8      CCGACCTCCAAGATTCGCTATGACAAT
Pfr7      CCGACCTCCAAGATTCGCTATGACAAT
Pbe5      CCGACCTCCAAGATTCGCTATGACAAT
Pma3      CCGACCTCCAAGATTCGCTATGACAAT
Pga11     CCGACCTCCAAGATTCGCTATGACAAT
Plo6      CCGACCTCCAAGATTCGCTATGACAAT
Pme2      CCGACCTCCAAGATTCGCTATGACAAT
Pre1      CCGACCTCCAAGATTCGCTATGACAAT
 11 27
Pfa4      ATGGTAGCGGATAACTGACTTCATCGA
Pvi10     ATGGTAGCGGATAACTGACTTCATCGA
Pcy9      ATGGTAGCGGATAACTGACTTCATCGA
Pkn8      ATGGTAGCGGATAACTGACTTCATCGA
Pfr7      ATGGTAGCGGATAACTGACTTCATCGA
Pma3      ATGGTAGCGGATAA.........etc
```

This file example .T7 was then submitted to the **phylip** program **dnapars** with the option multiple data sets indicating that there were 100 data sets to analyze, the first part of the output from this looked like this:

```
((R,((((M,K),L),N),Q),(J,P))),O)[0.0100];
((R,((((M,K),L),N),(J,Q)),P)),O)[0.0100];
((R,((((M,K),L),(J,N)),Q),P)),O)[0.0100];
((R,((((M,K),(J,L)),N),Q),P)),O)[0.0100];
((R,((((M,(J,K)),L),N),Q),P)),O)[0.0100];
((((((J,M),(R,K)),L),N),Q),P),O)[0.0100];
((((((J,(R,M)),K),L),N),Q),P),O)[0.0100];
(((((((R,J),M),K),L),N),Q),P),O)[0.0100];
((R,(((((J,M),K),L),N),Q),P)),O)[0.0100];
((((((R,(J,M)),K),L),N),Q),P),O)[0.0100];
(((R,J),((((M,K),L),N),Q),P)),O)[0.0100];
((J,(R,((((M,K),L),N),Q),P))),O)[0.0100];
((R,(J,((((M,K),L),N),Q),P))),O)[0.0100];
((R,((J,(((M,K),L),N),Q)),P)),O)[0.0100];
((R,(((J,((M,K),L),N)),Q),P)),O)[0.0100];
((R,((((J,((M,K),L)),N),Q),P)),O)[0.0100];
((R,(((((J,(M,K)),L),N),Q),P)),O)[0.0100];
(((J,(R,M)),(((K,L),N),Q),P)),O)[0.0100];
((((R,J),M),(((K,L),N),Q),P)),O)[0.0100];
(((R,(J,M)),(((K,L),N),Q),P)),O)[0.0100];
((M,((R,J),(((K,L),N),Q),P))),O)[0.0100];
(((R,J),(M,(((K,L),N),Q),P))),O)[0.0100];
(((R,J),((M,(((K,L),N),Q)),P)),O)[0.0100];
(((R,J),(((M,((K,L),N)),Q),P)),O)[0.0100];
(((R,J),((((M,(K,L)),N),Q),P)),O)[0.0100];
(((R,(M,(J,K))),(((L,N),Q),P)),O)[0.0100];
((((J,M),(R,K)),(((L,N),Q),P)),O)[0.0100];
(((R,((J,M),K)),(((L,N),Q),P)),O)[0.0100];
```

Notice at the end of each tree is associated a weight.

**Putting trees together: consensus.**

These trees are usually summarized by programs like phylip's consense. This has an output tree that looks like this:

```
Majority-rule and strict consensus tree program,
                               version 3.572c
CONSENSUS TREE:
the numbers at the forks indicate the number
of times the group consisting of the species
which are to the right of that fork occurred
among the trees, out of 100.00 trees
                      +----L
                  +-22.0
              +-22.0    +----K
              !   !
          +-22.0  +---------M
          !   !
          !   +-------------N
          !
  +100.0                +----O
  !   !           +-22.0
  !   !       +-22.0    +----R
  !   !       !   !
  !   +-22.0      +---------J
  !       !
  !       +-------------P
  !
  +---------------------Q

     remember: this is an unrooted tree!
```

**Proof of sparsity of the data.** Here are all the possible "boiled down" patterns in the `Vitis Vinifera` data:

```
11111111111111111111 11111111111111111121 11111111111111112111 11111111111111121111
                 187                    2                   17                    4
11111111111111211111 11111111111111213111 11111111111112111112 11111111111112112112
                   4                    1                    1                    1
11111111111231111113 11111111112111111211 11111111112111113211 11111111121111111131
                   1                    3                    1                    1
11111111121211111111 11111111121211111131 11111111122111121222 11111111211111111111
                   7                    1                    1                    5
11111111211111112111 11111111211111113111 11111111211111133111 11111111211111211111
                   1                    1                    1                    1
11111111211111331111 11111111221111111111 11111111221211322111 11111111221221112112
                   1                    1                    1                    1
11111111223121314352 11111111231311241111 11111111234311141441 11111112111111111111
                   1                    1                    1                    6
11111112111111112111 11111112111111113111 11111112112111113221 11111112345141141514
                   1                    1                    1                    1
11111121111111111111 11111121111111112111 11111121131311111111 11111123415141411114
                   3                    1                    1                    1
11111121111111113111 11111123111211111111 11111211111111111111 11111211311111111111
                   1                    1                    1                    1
11111211321111114111 11111221133132111132 11112111111111111111 11112111111111211111
                   1                    1                   10                    2
11112111111211111211 11112111111232131212 11112111113111111311 11112111131111111111
                   1                    1                    1                    1
11112111211111111111 11112111211111311111 11112111331111141111 11112113212121121121
                   1                    1                    1                    1
11112211111131113113 11112221111121111112 11112221111211113211 11112221131112121111
                   1                    1                    1                    1
11112221133332145362 11112221212121111111 11112221213132111111 11112311111111114111
                   1                    1                    1                    1
11121111111112111111 11121111113131111333 11121113423212444311 11121112133332113332
                   2                    1                    1                    1
11121112322121122111 11121221341111114115 11122111111112111112 11211111111111111111
                   1                    1                    1                    2
11211111111211113221 11211111121211111113 11211111122211111221 11211111311111111111
                   1                    1                    1                    1
11211113222121121211 11211211211111111111 11211221211211112221 11211321211111114111
                   1                    1                    1                    1
11212221211111111111 11212221211111112111 11212221211312112332 11212221213131111111
                   3                    1                    1                    1
11212221222122121212 11213111222221122231 11213331211331121321 11222222211111111111
                   1                    1                    1                    1
11222222222122212112 12111111111112111111 12111111111121112112 12113222212222111242
                   1                    1                    1                    1
12121111111112131112 12121112111111111111 12121112111112111111 12121112122121121121
                   1                    1                    1                    1
12121112122221221211 12121112122222222222 12121112312131211111 12121113111112114111
                   1                    1                    1                    1
12121332121411212411 12122222111212211222 12123112222222222222 12221111111112111111
                   1                    1                    1                    1
12221112122222222222 12221112222222222222 12221222111112111111 12221223212222222222
                   1                    1                    1                    1
12222112212221231221 12222222211113111111 12222222211122113112 12222222112112121211
                   1                    1                    1                    1
12222222113112111311 12222222121212122111 12222222212222222222 12222222222212112222
                   1                    1                    1                    1
12222222222212121221 12222222222222222212 12222222222222222222 12222222222222223234
                   1                    1                    1                    1
12222222222322233332 12222222345463455546 12222223442522162222 12222223454242426424
                   1                    1                    1                    1
12222232242124244214 12223222222222222222 12223222322332333223 12223332212121214111
                   1                    1                    1                    1
12223332241112141112 12223442325252533556 12232222211112124111 12232222111122111112
                   1                    1                    1                    1
12321121441412421111 12321123222222232222 12323332311411255441 12323332334243252623
                   1                    1                    1                    1
12324222312112112231 12324444444412144441 12324552464443278243 12333332211141211211
                   1                    1                    1                    1
12333343552513243261 12334333155553155565 12343353125222226562 12344435336352736615
                   1                    1                    1                    1
12345222121112252112
                   1
```

**How some regions are stable where others are are variable.**

| VVi | ISTGLGATLN | VAKPTKGSTV | AVFGLGAVGL | AAAEGARIAG | ASRIIGVDLN | PKRYEGAKKF |
|---|---|---|---|---|---|---|
| Zma1 | .......SI. | ....P..... | .......... | .......... | .......... | .S.F.E.R.. |
| Zma2 | .......... | ....A..... | .I........ | ..M...L... | ........I. | .AK..Q.... |
| Hvu1 | .......SI. | ....P..... | .I........ | .......... | .......... | AV.F.E.R.. |
| Hvu2 | .......... | .T..K..M.. | .I........ | ..M...MS.. | .......... | .AKH.Q.... |
| Hvu3 | L......... | ....K..... | .I........ | ..M...M... | .......... | .AK..Q.... |
| Tae | .......... | ....K..... | .I........ | ..M...M... | .......... | .AK..Q.... |
| Osa1 | ........I. | ....K-.... | .I........ | ........R. | ......I... | AN.F.E.R.. |
| Osa2 | F.SRF...V. | ....K..Q.. | .I........ | ..M...LS.. | .......... | .AKF.Q.... |
| Ath | L......... | ....K..QS. | .I........ | G......... | ........F. | S..FDQ..E. |
| Psa | .C......I. | ....KP..S. | .I........ | ........S. | .........V | SS.F.L.... |
| Fan | .......... | .R..K..... | .......... | .......M.. | .......... | SN.F.E.... |
| Tre | .C......V. | ....KP..S. | .I........ | .......MS. | .........V | SS.F.L.... |
| Stu | .......... | ........S. | .I........ | .......... | .......... | AS.F.Q.... |
| Pgl | .......SI. | ....P..... | .I........ | .......... | .......... | .S.F.E.... |
| Phy | .......... | .......... | .I........ | .......... | .......... | .S.FND.... |
| Pde | .......A.. | ....K..H.. | .......... | .......LS. | .......... | .S.FNE.... |
| Pta | V...M..... | ....K...S. | .I...G.... | .......... | ......I... | SD.F.K..L. |
| Fra | .......... | .R..K..... | .......... | .......M.. | .......... | SN.F.E.... |
| %Mal | .......... | ....K..... | .......... | .......LS. | .........H | SD.F.E.... |
| Lyc | .......S.. | ........S. | .I........ | .......... | .......... | AS.F.Q.... |

The regions of low variability have mostly dots in them, high variability is shown by the letters, (except for complete columns that indicate just a difference with the first taxa).

These patterns can be made even more visible by the use of color, (see Charnordic and Holmes, 1997).

**A consensus bootstrap tree for the vitis vinifera data.**

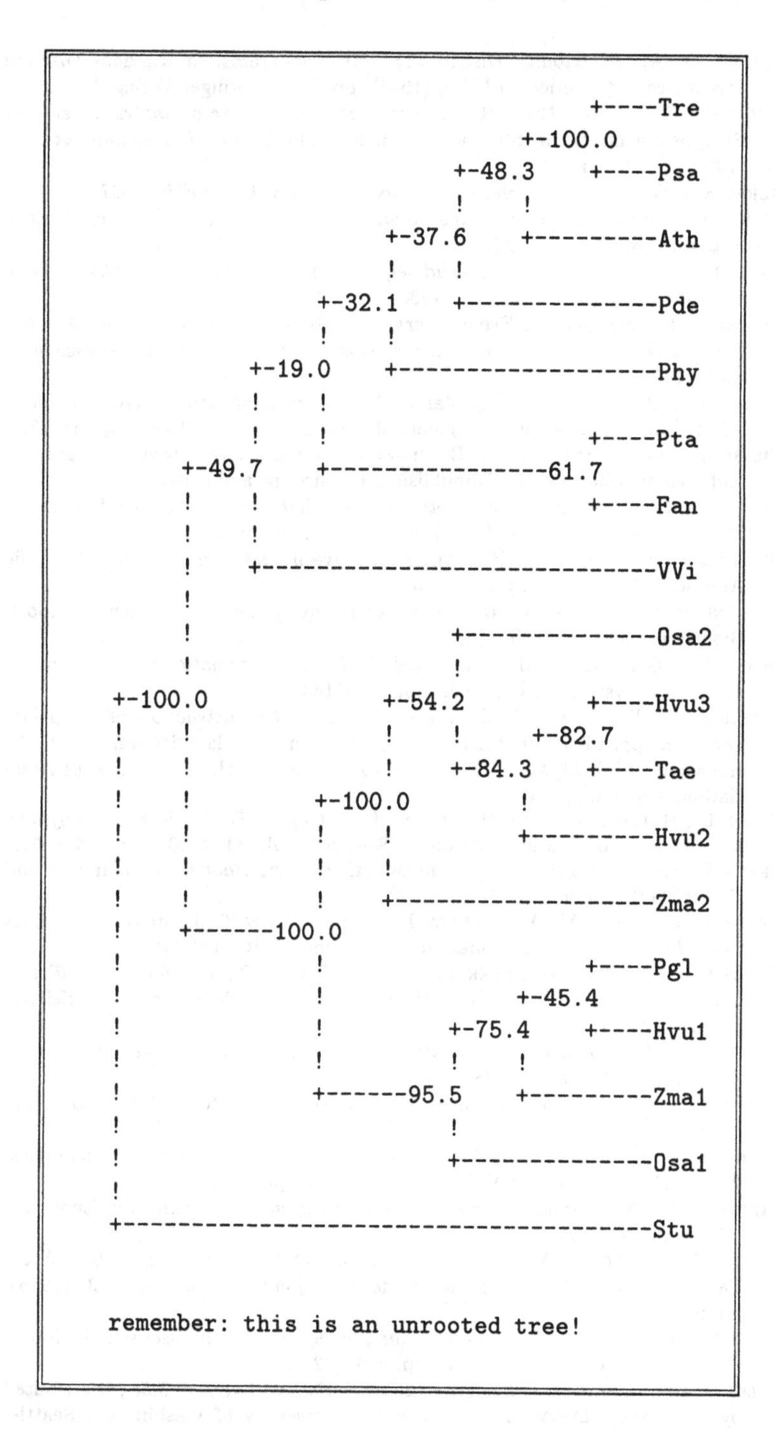

## REFERENCES


[1] ALDOUS D. A., *Probability Distributions on Cladograms*, in Random Discrete Structures, IMA series, **vol. 76**, (1996), pp. 1–18, Springer Verlag, NY.

[2] ANDERSON T. W. AND RUBIN H., *Statistical inference in factor analysis*, Berkeley Symposium on Math. Stat. and Probab., (Third), Ed. J. Neyman, **vol. 5**, (1956), pp. 111–150.

[3] BERRY AND GASCUEL O., *Strict Consensus Parsimony*, COCOON, 1997.

[4] BICKEL P. AND FREEDMAN D., *Some asymptotic theory for the Bootstrap*, Annalls Statistics, **9**, pp. 1196–1217.

[5] BREMER K., *The limits of amino-acid sequence data in angio-sperm phylogenetic reconstruction*, Evolution, **42**, (1988), pp. 795–803

[6] CHANG J., *Inconsistency of Evolutionary Tree Topology Reconstruction Methods when Substitution Rates Vary across Characters*, Mathematical Biosciences, **134**, (1996), pp. 189–215.

[7] CHANG J., *Full reconstruction of Markov Models on Evolutionary Trees: Identifiability and Consistency*, Mathematical Biosciences, **137**, (1996), pp. 51–73.

[8] CHARNOMORDIC B. AND HOLMES B., `Dnaview`, an interactive viewer for alignment and tree building, (1997), Unpublished manuscript and software.

[9] CHERNOFF H., *Problems with Bootstrapping Phylogenies*, IMA conference on Statistics and Genetetics, (1997), unpublished communication.

[10] DIACONIS P. AND EFRON B., *Computer intensive methods in statistics*, Scientific American, **248**, (1983), pp. 116–130

[11] DIACONIS P. AND HOLMES S., *Random walks on phylogenetic trees*, Techical report, Biometrics Unit, Cornell, (1997).

[12] DOYLE J.J., *Gene trees and species trees: Molecular systematics as one-character taxonomy*, Syst. Bot., **17**, (1992), pp. 144–163.

[13] EDWARDS, A. W. F. AND L. L. CAVALLI-SFORZA, *Reconstruction of evolutionary trees*, pp. 67–76, in Phenetic and Phylogenetic Classification, ed. V. H. Heywood and J. McNeill., Systematics Association **vol. 6**, Systematics Association, London, (1964).

[14] EFRON B., HALLORAN E. AND HOLMES S., *Bootstrap confidence levels for phylogenetic trees*, Proc. National Academy Sciences, **vol. 93**,(1996), pp. 13429–34.

[15] EFRON B. AND TIBSHIRANI R., *An Introduction to the Bootstrap*, Chapman and Hall, (1993), London.

[16] ERDÖS P. L., STEEL M. A., SZÈKELY L., AND WARNOW T. J, *Inferring big trees from short sequences*, to appear in Proceedings of ICALP, (1997).

[17] ERDÖS P. L., STEEL M. A., SZÈKELY L., AND WARNOW T. J, *A few logs suffice to build (almost) all trees*, (I) and (II) Tech reports, U. Penn. Computer Science Dept, (1997).

[18] FARRIS J.S., *A successive approximations approach to character weighting*, Syst. Zool, **18**, (1969), pp. 374–385.

[19] FARRIS J. S., *Methods for computing Wagner trees*, Syst. Zool., **219**, (1970), pp. 83–92.

[20] FARRIS J. S., *The logical basis of phylogenetic analysis*, in Advances in cladistics, **vol. 2**, (N. Platnick and V. Funk, eds.) (1983), pp. 7–36.

[21] FARRIS J. S., *The information content of the phylogenetic system*, Syst. Zool., **28**, (1979), pp. 483–519.

[22] FARRIS, J. S., ALBERT, V. A., KÄLLERSJO, M., LIPSCOMB,D. AND KLUGE A.G., *Parsimony jackknifing outperforms neighbor-joining*, Cladistics, **12**, (1996), pp. 99–124.

[23] FELSENSTEIN, J., *Statistical inference of phylogenies (with discussion)*, Journ. Royal Stat. Soc. A, **146**, (1983), pp. 246–272.

[24] FELSENSTEIN, J., `PHYLIP`, *(Phylogeny Inference Package) version 3.5c.*, Distributed by the author. Department of Genetics, University of Washington, Seattle, (1993).

`http://evolution.genetics.washington.edu/phylip.html`

[25] FOULDS L. R. AND GRAHAM R. L., (1982) *The Steiner tree problem in phylogeny is NP-complete*, Adv. Appl. Math., **3**, (1982), pp. 43–49.

[26] FREEDMAN D. AND LANE D., *Significance testing in a non stochastic setting*, Festschrift for Eric Lehmann, (1983), pp. 185–208.

[27] FREEDMAN D. AND PETERS S. C., *Some notes on the bootstrap in regression problems*, Journ. Bus. Ec. St., **2**, (1984), pp. 406–409.

[28] FRIEDMAN J. H. AND RAFSKY L., *Multivariate generalizations of the Wald-Wolfowitz and Smirnov two-sample tests*, Annals Statistics, **7**, (1979), pp. 697–717.

[29] GARDNER M., *The Last Recreations*, Copernicus-Springer Verlag, NY, (1997).

[30] GOLOBOFF P., *Nona*, available from J. Carpenter, Entomology Dept, American Museum of NAtural History , 79th st.,New York, NY 10024-5192 (1995).

[31] GOLOBOFF P., *Self-weighted optimization: Tree searches and character state reconstructions under implied transformation costs*, Cladistics, **12**, (1997), pp. 225–246.

[32] HARSHMAN J., *The effect of irrelevant characters on bootstrap values*, Syst. Biol., **43**, (1994), pp. 419–424.

[33] HARVEY P. H. AND PAGEL M. D., *The comparative method in Evolutionary Biology*, Oxford University Press, Oxford, (1991).

[34] HILLIS D. M., BULL J. J. WHITE M. E., BADGETT M. R. AND MOLINEUX I. J., *Experimental Phylogenies: generation of a known phylogeny*, Science 255, (1992), pp. 589–592.

[35] HILLIS D., MORITZ C., AND MABLE B.,, *Molecular Systematics*, Sinauer, (1996), Boston.

[36] KLUGE, A. C. AND FARRIS J. S., *Quantitative phylogenetics and the evolution of anurans*, Syst. Zool., **18**, (1969), pp.1–32.

[37] LAVIT, C., ESCOUFIER, Y., SABATIER, R., AND TRAISSAC, P., *The ACT (STATIS method)*, Comput. Statist. Data Analysis, **18**, (1994), pp. 97–119.

[38] LI W. H., *Molecular Evolution*, Sinauer, Boston, (1997).

[39] LI S., PEARL D. K., DOSS H., *Phylogenetic Tree Construction using MCMC*, Technical report no 583. Ohio Statistics Dept., (1996), submitted to Journ. American Statistical Association.

[40] MAU, B., NEWTON, M. A., AND LARGET B., *Bayesian phylogenetic inference via Markov Chain Monte Carlo Methods*, (1999) to appear Biometrics, vol.55.

[41] NEWTON, M. A., *Bootstrapping Phylogenies: Large deviations and dispersion effects*, Biometrika, **83**, (1996), pp. 315–328.

[42] PAGE R. D. AND CHARLESTON M., *From gene to Organismal Phylogeny: Reconciled Trees and the Gene Tree/Species Tree Problem*, (1997), Tech rep. Univ.Glascow., `http://taxonomy.zoology.gla.ac.uk/rod/pubs.html`

[43] RAMBAUT, A. AND GRASSLY, N. C., *Seq-Gen: An application for the Monte Carlo simulation of DNA sequence evolution along phylogenetic trees*, Comput. Appl. Biosci., **13**, (1997), pp. 235–238.

[44] RICE K., STEEL M., WARNOW T. AND YOOSEPH S., *Getting better topology estimates of difficult evolutionary trees*, U. Penn. Computer Science, Tech. Report, (1997).

[45] SANDERSON M., *Objections to bootstrapping phylogenies: a critique*, Syst. Biol., **44**, (1995), pp. 299–320.

[46] SARNI-MANCHADO, P., VERRIÈS C. AND TESNIÈRE C., *Molecular characterization and structural analysis of one dehydrogenase gene (GV-adh 1) expressed during ripening of grapevine (Vitis vinifera L.) berry*, Plant Science, **125**, (1997), pp. 177–187.

[47] SATTAH S. AND TVERSKY A., *Additive similarity trees*, Psychometrika **vol 42 no 3**,(1977), pp. 319–345.

[48] SCHRÖDER E., *Vir Combinatorische Probleme*, Zeit. Pur. Math. Phys., **vol 15**, (1870), pp. 361–376.

[49] STANLEY R., *Enumerative Combinatorics*, **vol I**, 2nd edition (1996).

[50] STRIMMER, K. AND VON HAESELER, A., *Quartet Puzzling: a quartet maximum likelihood method for reconstructing tree topologies*, Mol. Biol. Evol., **13**, pp. 964–969.

[51] SWOFFORD, PAUP 4.0, (1998), Available from Sinauer,Boston.

[52] TUFFLEY AND STEEL M., (1997) *Links between Maximum Likelihood and Maximum Parsimony under a simple model of substitution*, Technical Report.

[53] WATERMAN M. S. AND SMITH T. F., *On the similarity of dendograms*, Jour. Theoret. Biology, **73**, (1978), pp. 789–800.

[54] WHEELER W., *MALIGN*, Dept. of invert., Am. Museum of Natural History, NY.

[55] YOKOYAMA S. AND HARRY D. E., *Molecular Phylogeny and evolutionary rates of alcohol dehydrogenases in verterbrates and plants*, Mol. Biol. Evol., **10**, (1993), pp. 1215–1226.

[56] ZHARKIKH, A. AND LI W. H., *Estimation of confidence in phylogeny: The complete and partial bootstrap technique*, Mol. Phylogenet. Evol., **4**, (1995), pp. 44–63.

# DETECTING LINKED GENOMIC MUTATIONS*

FRANCOISE SEILLIER-MOISEIWITSCH†, RONALD SWANSTROM‡, AND MICHAEL Z. MAN§

**Abstract.** Detecting whether mutations at different positions are linked may provide insights into the protein structure encoded in a genome as well as into the molecular evolution of this genome. Statistical methodology applied so far pertains to testing the statistical independence of two positions. We present some statistical procedures specifically developed to detect linkage and excess or scarcity of double mutations. One test conditions on the number of double mutations and is akin to a $\chi^2$-test for goodness of fit to a specified multinomial distribution. The other, in the form of a binomial test, checks that the number of double mutations is as expected under the assumption of independence. Linked mutations may be the result of independent substitutions passed on together during replication. We construct tests for mutational independence under the observed phylogeny, which attempts to remove the phylogenetic effect from the observed linkages. This methodology is implemented on amino-acid sequences from the V3 loop of the gene coding for the envelope protein of the human immunodeficiency virus.

**Key words.** Test of independence, Phylogeny, Amino-acid sequences.

**AMS(MOS) subject classifications.** Primary 62F03, 92D20; Secondary 6207, 62F05, 62F30, 62P10, 92D10, 92D15, 92D20.

**1. Description of the problem.** Genomes such as that of the human immunodeficiency virus (HIV) evolve rapidly. The inefficiency of the replication process in HIV, like in any retrovirus, gives rise to many variants. In their genetic material, these viral isolates vary considerably from individual host to individual host as do, to a lesser extent, those sampled from a single individual. The observed variability reflects both the viability of the mutants and selection pressures from the immune system. For a comprehensive description of this variability for HIV and a review of the statistical methods utilized to analyze it, see [11].

Within a given set of sequences there most frequently is a favoured amino acid at each position. The string of these predominant amino acids can be considered a *consensus sequence.* Some positions show only little or no substitution. At other sites, there is substantial heterogeneity, suggesting only limited sequence constraints. Sometimes, a small number of amino

*The authors are indebted to L. Milich for the preparation of the data set and to M. Daley, G. Hafley, J. Sgalla and M. Karnoub for their programming skills.

†Department of Biostatistics, University of North Carolina, Chapel Hill, North Carolina 27599-7400. Her work was supported in part by NSF grant DMS-9305588, AmFAR grant 70428-15-RF and NIH grant R29-GM49804.

‡Department of Biochemistry and Biophysics and Lineberger Comprehensive Cancer Center, University of North Carolina, Chapel Hill, North Carolina 27599-7295. His research is funded in part by NIH grant NIAID R01-AI253210.

§Department of Biostatistics, University of North Carolina, Chapel Hill, North Carolina 27599-7400. He is the recipient of a Pfizer postdoctoral fellowship.

acids appear with near equal frequency, implying near equivalency of function. These *polymorphisms* at one position are usually associated with the consensus amino acid at other positions. Finally, at other positions, some otherwise deleterious changes may be rescued by a compensatory mutation, allowing both departures from consensus to remain in the population only through their linked appearance and thus maintaining structure (and therefore viability) or even conferring a survival advantage to these variants. There are a number of instances where it is important to identify these double substitutions.

Detecting linked mutations may indeed provide insights into the protein structure encoded in a genome as well as into the evolution of this genome. The proteins making up a virus like HIV cannot all be observed directly. Their structures often need to be inferred. Genomic interaction is likely to be caused by spacial proximity.

Most methods for reconstructing *phylogenetic trees* rely on the assumption of independence among positions [5, 12]. If this is violated, the computed distances (usually the number of positions at which two sequences differ) are overestimates of the actual distances. This results in unreliable estimates of the tree configuration and of its branch lengths. It is therefore important to evaluate the suitability of the assumption of independence on the data of interest.

*Signature pattern analysis* is applied to the problems of determining epidemiological linkage (whether variants from a single original source are present in different individuals) and superinfection (whether an individual is infected by strains from different origins) [8]. A *signature pattern* refers to a set of non-contiguous amino acids or nucleotides that distinguish most effectively a *query* set of variable sequences from a *background* or *control* set. For epidemiological linkage, as in the study focussing on an HIV-positive Florida dentist [9], one searches for highly similar viral strains in different individuals to assess the possibility of transmission, while for superinfection one needs to demonstrate the presence of very different quasispecies in an individual in order to conclude that this individual was infected by two separate strains. Clearly, in estimating the significance levels of observed similarities it is important to take into account associations of mutations.

We review the methodology to date in Section 2 and present the new tests in Sections 3 and 4. An illustrative data analysis is presented in Section 5. Section 6 contains a closing discussion.

## 2. Previous studies.

Statistical methodology applied so far pertains to testing the statistical independence between two positions [7, 2].

### 2.1. Information-based measures.

To detect covariation among mutations, [7] utilizes an information-theoretic measure computed on the whole data set. In the present context, they are helpful in comparing the amino acid (or nucleotide) distributions at different sites along the genome. Let $i$ denote a position along the genome and $a$ an amino acid appearing at that position. If $p_i(a)$ represents the frequency of appearance of amino acid $a$ at position $i$ and is usually estimated from the data), then one defines the *Shannon entropy*

$$H(i) = -\sum_{a_i} p_i(a_i)\ log\, p_i(a_i) \tag{2.1}$$

where $a_i$ refers to amino acids appearing at position $i$. This is a measure of the variability at a single position. This concept easily generalizes to quantify variability at two positions, $i$ and $j$, simultaneously:

$$H(i,j) = -\sum_{a_i,b_j} p_{ij}(a_i,b_j)\ log\, p_{ij}(a_i,b_j)\ . \tag{2.2}$$

From these, one can construct a measure of covariation, the *mutual information*, defined as

$$M(i,j) = H(i) + H(j) - H(i,j)\ . \tag{2.3}$$

This function takes its minimum value of 0 when either there is no variation or the positions $i$ and $j$ vary independently. It attains its maximum when the same pairings always occur. $M(i,j)$ in fact denotes the likelihood-ratio statistic for testing the hypothesis of independence between two positions against that of arbitrary dependence:

$$M(i,j) = \sum_{a_i,b_j} p_{ij}(a_i,b_j)\ log\,\{p_{ij}(a_i,b_j)/p_i(a_i)\,p_j(b_j)\}\ . \tag{2.4}$$

To focus on the covariability of a specific pair of residues, define $K_{i,j}(a,b)$ as the $M(i,j)$ statistic obtained by replacing the 20-letter alphabet (i.e., the amino acids) at site $i$ by $\{a,\overline{a}\}$ (i.e., $a$ and not $a$) and at site $j$ by $\{b,\overline{b}\}$ [2]. $K_{i,j}(a,b)$ is therefore the likelihood-ratio statistic for testing the assumption of independence

$$H_0 : p_{ij}(x,y) = p_i(x)p_j(y) \text{ for all x, y}$$

against the alternative hypothesis of quasi-independence [1]

$$H_1 : p_{ij}(x,y) = f_i(x)g_j(y) \text{ for some functions } f_i \text{ and } g_j \text{ and } x \neq a \text{ and } y \neq b.$$

Finally,

$$K(i,j) = \max_{a,b} K_{ij}(a,b) \tag{2.5}$$

is reportedly suited to identify linked mutations.

To obtain the reference distribution for the statistics $M(i,j)$, the amino acids at each position are permutated independently. A large number $B$ (here 100,000) of pseudo-data sets are thus created. From each of these pseudo-data sets, the statistics are computed for all pairs of positions. The number $b_{ij}$ of pseudo-statistics for the pair $(i,j)$ that exceed the observed $M(i,j)$ is computed for all $(i,j)$, which leads to the observed significance level $(b_{ij}+1)/(B+1)$. When $b_{ij}=0$, the observed significance is thus $(B+1)^{-1}$. Similarly for $K(i,j)$.

**2.2. Measure of association.** [2] also uses the statistic developed by [6]:

$$G(i,j) = \frac{1}{2}\{\sum_a \hat{p}_{ij}(a,max) + \sum_b \hat{p}_{ij}(max,b) - \hat{p}_i(max) - \hat{p}_j(max)\} \Big/ \{1 - \frac{1}{2}(\hat{p}_i(max) + \hat{p}_j(max))\} \tag{2.6}$$

where

$$\hat{p}_{ij}(a,max) = \max_b \hat{p}_{ij}(a,b), \quad \hat{p}_{ij}(max,b) = \max_a \hat{p}_{ij}(a,b)$$

and

$$\hat{p}_i(max) = \max_a \hat{p}_i(a) \ .$$

$G(i,j)$ represents the relative reduction in the probability of guessing incorrectly the amino acid at site $j$ associated with the knowledge of the residue at site $i$ and vice versa. If the two positions are not associated $G(i,j) = 0$ . The reference distribution of $G(i,j)$ is generated through the process described in the previous section.

**3. Tests of statistical independence of mutations.** Let us describe the various amino acids present at two positions along the genome across a set of DNA sequences via a contingency table. The rows of this table represent the amino acids in the first position and the columns those in the second position. For the sake of simplicity, let us put in the first cell the number of sequences that harbour the *consensus* pair (i.e. that with the highest frequency). Then, the first row and column (the *outer table*, say) contain sequences with a single departure from the consensus pair (in the first or second position but not both) and represent the *observed polymorphisms.* The rest of the table (the *inner table*) deals with double mutations.

TABLE 1
*Probability model for independent mutations.*

| | | Position $i$ | | | |
|---|---|---|---|---|---|
| | | A | C | D | E |
| Position $j$ | S | $(1-\alpha)(1-\beta)$ | $\alpha_1\ (1-\beta)$ | $\alpha_2\ (1-\beta)$ | $\alpha_3\ (1-\beta)$ |
| | T | $(1-\alpha)\ \beta_1$ | $\alpha_1\ \beta_1$ | $\alpha_2\ \beta_1$ | $\alpha_3\ \beta_1$ |
| | V | $(1-\alpha)\ \beta_2$ | $\alpha_1\ \beta_2$ | $\alpha_2\ \beta_2$ | $\alpha_3\ \beta_2$ |
| | W | $(1-\alpha)\ \beta_3$ | $\alpha_1\ \beta_3$ | $\alpha_2\ \beta_3$ | $\alpha_3\ \beta_3$ |
| | Y | $(1-\alpha)\ \beta_4$ | $\alpha_1\ \beta_4$ | $\alpha_2\ \beta_4$ | $\alpha_3\ \beta_4$ |
| | | polymorphisms outer table | | double mutations inner table | |

Let $N$ denote the total number of sequences, $n$ the number of sequences with double mutations and $n_{ij}$ the count in cell $(i,j)$. $\alpha$ represents the probability of mutation away from the consensus at position $i$ and $\beta$ at position $j$:

$$\alpha = \sum_{u=1}^{c-1} \alpha_u \quad \text{and} \quad \beta = \sum_{v=1}^{r-1} \beta_v \tag{3.1}$$

where $\alpha_u$ and $\beta_v$ denote the probabilities of particular substitutions and $c$ is the number of different amino acids occurring at position $i$ and $r$ the corresponding number at position $j$ (Table 1).

To detect mutational linkage one should therefore focus on the inner table (i.e. the whole table barring the first row and column) and determine whether the hypothesis of statistical independence holds in this limited context. The inner-table cells thus enter into the test statistic while the whole table (or the outer table only) is utilized for estimation of the parameters. Furthermore, the biological process that generates the observations in the outer table is subject to different constraints to the process that gives rise to the data in the inner table. Consequently, one should avoid conditioning on summary statistics that amalgamate these two sources. For these reasons, we do not make use of Fisher's exact test or its extensions.

**3.1. Conditioning on the total number of double mutations.** First, we condition on the total number of sequences falling into the inner table and check whether the observed distribution across the individual cells follows the assumption of independence. The test statistic of choice is the $\chi^2$-statistic

$$\sum_{i,j \,\in\, inner\ table} \frac{(O_{ij} - E_{ij})^2}{E_{ij}}$$

where $O_{ij}$ and $E_{ij}$ are, respectively, the observed and the expected number of sequences in cell $(i, j)$. $E_{ij}$ is calculated on the basis of the hypothesis of statistical independence.

Now we turn to the choice of estimated frequencies to be entered in these expectations. Those based on the marginal totals are commonly used, i.e.

$$E_{ij} = n \frac{n_{i.}\; n_{.j}}{\sum_{i=2}^{r} \sum_{j=2}^{c} n_{i.}\; n_{.j}} \tag{3.2}$$

where $n_{i.} = \sum_{j=1}^{c} n_{ij}$ and $n_{.j} = \sum_{i=1}^{r} n_{ij}$. Here, because these totals include the frequencies of double mutations and in view of the possibility that the biological process generating joint mutations may be different from that causing single mutations at a pair of positions, it may be more reliable to use the relative frequencies in the first row and column (normalized so that each vector - i.e. row or column - sums to one), i.e.

$$E_{ij} = n \frac{\hat{p}_{i1}\,\hat{p}_{1j}}{\sum_{i=2}^{r}\sum_{j=2}^{c}\hat{p}_{i1}\,\hat{p}_{1j}} \tag{3.3}$$

where $\hat{p}_{i1} = n_{i1}/n_{.1}$ and $\hat{p}_{1j} = n_{1j}/n_{1.}$. As demonstrated in Table 2, the usual reference distribution for the $\chi^2$-statistics is inappropriate for moderate sample sizes. The entries of Table 2 represent the number of test statistics (out of 1,000) falling above the 90th, 95th and 99th percentiles of the $\chi^2_3$-distribution. The underlying probability distributions for the rows and columns of the simulated tables are (.70,.15,.15). The cell counts were generated under the assumption of independence. Other choices for the underlying distributions yielded similar outcomes.

TABLE 2

*Chi-square test (* = outside 2 SD's , ** = outside 3 SD's from the expected value).*

| | Percentiles | | |
|---|---|---|---|
| Sample Size | 90 | 95 | 99 |
| Marginal Totals | | | |
| 100 | 46 ** | 21 ** | 3 * |
| 200 | 41 ** | 20 ** | 1 * |
| 400 | 54 ** | 26 ** | 4 |
| Outer-Table Frequencies | | | |
| 100 | 212 ** | 142 ** | 59 ** |
| 200 | 196 ** | 122 ** | 51 ** |
| 400 | 182 ** | 110 ** | 39 ** |

Therefore, just as [7] and [2] resorted to permuting the original date to assess the the significance of their test statistics, we need to resample from our original distribution to construct relevant reference distributions. However, here, it is done parametrically: the parameters of the original distribution are estimated and it is from this estimated distribution that the resampling is performed. We utilize the $BC_a$ method with 500 bootstrap samples [4]. We need to modify the usual procedure as we are interested in hypothesis testing. Indeed, the observed contingency table may not reflect statistical independence between rows and columns. We therefore first simulate a contingency table, under the assumption of independence, utilizing estimates of row and column probabilities from the data. We then generate bootstrap samples (parametrically) from this table. We compute the test statistic from each of these samples and the resulting empirical distribution serves as our reference distribution for the test statistic calculated on the original data. For the $\chi^2$-statistics, the bootstrap version demonstrates good tail properties for moderate sample sizes whatever the estimation procedure (Table 3).

TABLE 3
*Bootstrap chi-square test.*

| | Percentiles | | |
|---|---|---|---|
| Sample Size | 90 | 95 | 99 |
| Marginal Totals | | | |
| 100 | 110 | 59 | 4 |
| 200 | 97 | 55 | 7 |
| Outer-Table Frequencies | | | |
| 100 | 92 | 49 | 9 |
| 200 | 96 | 51 | 12 |

When a contingency table is sparse, some of the rows and/or columns with low counts are aggregated. To ensure that the statistical significance of the test statistic is driven by the non-aggregated cells, we decompose the overall statistic into residuals for the individual cells. We are specifically interested in the sum of such residuals over the non-aggregated cells in the inner table (*inner-cell statistics*).

**3.2. Total number of double mutations.** While the null hypothesis of statistical independence may hold within the inner table, the total number of double mutations may be inflated or suppressed. This would indicate that different biological processes are operating in the inner and outer tables. To assess this eventuality, we apply a binomial test on the aggregated inner table:

$$\frac{X - N\hat{p}}{\sqrt{N\hat{p}\,(1-\hat{p})}} \tag{3.4}$$

where $X$ denotes the observed number in the inner table, $N$ the total number of sequences considered and $p = \alpha\beta$ the probability of a double mutation assuming independence among positions. Again, $\hat{p}$ is computed from either the entries in the outer table

$$\hat{p} = \frac{n_{1.} - n_{11}}{n_{1.}} \, \frac{n_{.1} - n_{11}}{n_{.1}} \tag{3.5}$$

or the marginal totals

$$\hat{p} = \frac{N - n_{1.}}{N} \, \frac{N - n_{.1}}{N} \,. \tag{3.6}$$

Again, the usual reference distribution for this test statistic is inadequate in this context. The entries of Table 4 represent the number of test statistics, out of 1,000, falling below the 1st, 5th and 10th percentiles and above the 90th, 95th and 99th percentiles of the normal distribution. The same probabilistic set-up as above is utilized for the simulations. A bootstrap procedure similar to that described previously gives rise to a binomial test statistic with adequate sampling properties when one estimates $p$ from the marginal totals even for low sample sizes. When $\hat{p}$ is based on the polymorphism frequencies, the bootstrap does not perform as well for all combinations of row and column probabilities.

TABLE 4
*Binomial test (* = outside 2 SD's , ** = outside 3 SD's).*

| | Percentiles | | | | | |
|---|---|---|---|---|---|---|
| Sample Size | 1 | 5 | 10 | 90 | 95 | 99 |
| Marginal Totals | | | | | | |
| 100 | 1 * | 8 ** | 38 ** | 43 ** | 11 ** | 1 * |
| 200 | 1 * | 13 ** | 44 ** | 48 ** | 14 ** | 0 ** |
| 400 | 0 ** | 7 ** | 35 ** | 37 ** | 13 ** | 0 ** |
| Outer-Table Frequencies | | | | | | |
| 100 | 23 ** | 99 ** | 155 ** | 179 ** | 139 ** | 70 ** |
| 200 | 39 ** | 105 ** | 169 ** | 200 ** | 136 ** | 61 ** |
| 400 | 27 ** | 97 ** | 172 ** | 170 ** | 114 ** | 51 ** |

TABLE 5
*Bootstrap binomial test.*

| | Percentiles | | | | | |
|---|---|---|---|---|---|---|
| Sample Size | 1 | 5 | 10 | 90 | 95 | 99 |
| Marginal Totals | | | | | | |
| 100 | 7 | 54 | 115 | 104 | 46 | 9 |
| 200 | 8 | 51 | 93 | 107 | 50 | 3 * |
| Outer-Table Frequencies | | | | | | |
| 100 | 8 | 50 | 107 | 95 | 47 | 4 |
| 200 | 9 | 46 | 102 | 110 | 52 | 10 |

**4. Linkage or phylogenetic relationships?** Correlated mutations may be the result of independent substitutions being passed on together during replication; hence, the statistical significance would be caused by the phylogenetic relationship among the viral sequences. We need to find out whether these double mutations are due to parallel evolution or the result of a random occurrence of two independent mutations early in the replication process (so that all sequences harboring these substitutions are located on the same, albeit large, branch of the tree). We therefore developed some testing procedures that remove the phylogenetic effect from the observed linkages.

We cannot simply reconstruct the tree in order to investigate this issue: most methods for estimating phylogenies involve the assumption of independence among positions, others (such as the likelihood method and the method of invariants) cannot deal with such a large data set. We resort to reproducing the evolution of the sequences and generate the reference distribution for the test statistic on the basis of the simulated sequences.

From the sequence data, for each position, we estimate the frequency distribution of the amino acids. In view of the impossibility of some amino acids at specific positions, it is not reasonable to pool the estimates for different positions into overall estimates. We take the observed frequencies as substitution rates since they reflect structural constraints and immune selection pressures.

The event of a mutation at a specific site is modelled as a two-step process. First, whether the position undergoes a change is governed by the overall rate for the genome under study. In the context of HIV sequences, this is the error rate for reverse transcriptase, i.e. .0005 per site per replication [10]. Then, in case of mutation, the specific substitution is governed by the transition matrix described above.

The simulation starts with the consensus sequence as seed. It is subjected to the mutation process a random number of times. We set this number between 100 and 2,400. Indeed, for HIV, mutations occur at the time of replication. Its replication rate is approximately 240/year and this number represents the number of replications before transmission. This sequence thus gives rise to offspring sequences (*branches*). In our context, the branching process reflects HIV transmission: no offspring with probability .20 and 1 to 5 offpring each with probability .16 [3]. The tree is grown by repeating this process many times (with the output sequences from the previous generation as seeds). We obtain a total of 10,000 to 20,000 sequences. We sample without replacement as many sequences as there are in the original data set. From these we compute the test statistic. This sampling is performed a large number of times (here 1,000) to build up a reference distribution.

**5. Data analysis.** To illustrate the methodology, we selected a set of 356 HIV sequences. Each contains the 35 amino acids that form the V3 loop of the gene coding for the envelope protein. These sequences are independent in that each is sampled from a different individual. However, there is no guarantee that they are not epidemiologically related. They are chosen for their homogeneity with respect to selection pressures and origin. They are indeed all from the macrophage-tropic strain and from the so-called B clade (covering North America and Western Europe), which implies that they have the same consensus sequence. We elected to discard tables with all inner cells containing fewer than 4 observations, as these cannot lead to results of scientific interest.

The entries of Table 6 represent the linkages giving rise to statistically significant inner-cell $\chi^2$-test statistics (at least at the 99% level for one-sided tests). We indicate whether the observed test statistic for the cell falls above the 99.5 percentile (*), the 99.9 percentile (**) and the maximum (***) of the reference distribution, computed under the assumption of statistical independence among the positions. The reference distribution is generated in two ways: by using the bootstrap procedure and by simulating the molecular evolution of the sequences. The reported results are the least significant of both set-ups. For the estimation procedure relying on the polymorphism information, in order to avoid infinite residuals, whenever a cell $(i, j)$ in the inner table is non-zero, we replace zeroes in the cells $(1, j)$ and $(i, 1)$ by ones.

TABLE 6
*Chi-square tests for the inner cells (overall test statistics significant). The numbers in parentheses denote the number of sequences showing these linkages.*

| Positions | Substitutions | Marginals . | Polymorphisms |
|---|---|---|---|
| 5, 27 | $N \to S, I \to T$ (13) | | 75.08 *** |
| 13, 18 | $H \to P$ , $R \to Q$ (5) | 13.42 ** | |
| | $H \to P$ , $R \to K$ (7) | | |
| 13, 32 | $H \to P$ , $Q \to K$ (6) | 12.08 *** | |
| 14, 18 | $I \to L$ , $R \to Q$ (11) | 30.49 * | 124.63 *** |
| 14, 20 | $I \to L$ , $F \to W$ (14) | 26.35 *** | 71.85 *** |
| 14, 25 | $I \to L$ , $E \to Q$ (12) | 30.83 *** | |
| | $I \to M$ , $E \to D$ (19) | | |
| | $I \to L$ , $E \to D$ (5) | | |
| 15, 22 | $G \to A$ , $T \to A$ (12) | 0.07 * | 0.08 * |
| 16, 29 | $P \to W$ , $D \to N$ (5) | 8.57 ** | 18.98 ** |
| 18, 20 | $R \to Q$ , $F \to W$ (11) | 39.38 *** | 295.54 *** |
| 18, 22 | $R \to S$ , $T \to A$ (9) | | 46.30 *** |
| | $R \to Q$ , $T \to A$ (4) | | |
| | $R \to K$ , $T \to A$ (13) | | |
| 18, 25 | $R \to Q$ , $E \to Q$ (9) | 52.67 ** | 215.58 *** |
| | $R \to K$ , $E \to D$ (5) | | 3.01 * |
| 20, 25 | $F \to W$ , $E \to Q$ (11) | 42.82 *** | 108.42 *** |

Table 7 contains the results of the binomial tests. Again, we indicate whether the observed test statistic falls above the 99.5 percentile or below the 0.5 percentile (*), above the 99.9 percentile or below the 0.1 percentile (**), and above the maximum or below the minimum (***) of the reference distribution. Table 8 shows the results for the binomial and $\chi^2$-tests for those pairs with joint significance of K, M and G at the .00001 level, the level utilized by [2]. 100,000 permutations were performed to construct the reference distributions for these statistics.

TABLE 7
*Binomial tests.*

| Positions | Marginal Information | Polymorphism Information |
|---|---|---|
| 2, 13 | 2.85 * | 6.30 ** |
| 5, 16 | 2.92 ** | 4.48 *** |
| 5, 27 | 5.60 *** | 10.90 *** |
| 10, 25 | | 5.29 * |
| 11, 13 | -3.95 *** | -5.92 *** |
| 13, 25 | 2.43 *** | 7.59 *** |
| 14, 16 | 8.04 *** | 34.83 *** |
| 14, 18 | 5.30 *** | 11.00 *** |
| 14, 19 | 3.08 * | 5.83 * |
| 14, 20 | 5.21 *** | 11.19 *** |
| 15, 22 | 1.98 * | 6.12 * |
| 16, 22 | 1.84 * | 4.61 * |
| 18, 20 | 5.49 *** | 10.16 *** |
| 32, 34 | 4.42 *** | 6.64 *** |

TABLE 8
***Results of the binomial and chi-square tests for those positions which show a joint significance of K, M and G at the .00001 level. — indicates that the table has no inner cell with a count greater than 4.***

| Positions | Binomial Tests | Chi-square Tests |
|---|---|---|
| 5 , 27 | *** | *** |
| 7 , 30 | — | — |
| 8 , 27 | — | — |
| 11 , 33 | — | — |
| 14 , 16 | *** | |
| 14 , 18 | *** | *** |
| 14 , 20 | *** | *** |
| 18 , 20 | *** | *** |
| 29 , 33 | — | — |
| 30 , 31 | — | — |

**6. Discussion.** In our data analyses, the two estimation procedures yield, more often than not, the same inference. In the sensitivity analysis, the estimation procedure employing the polymorphism information fairs better that that relying on the marginal totals for both the binomial and the $\chi^2$-statistics. It is interesting to note that, for both tests, the weaker linkages (with respect to the bootstrap reference distribution) disappear when the reference distribution reflects the evolutionary process.

The pairs (7, 30), (8, 27), (11, 33), (29, 33), (30, 31) all yield observed K, M and G statistics significant at the .00001 level. Yet, the data for these pairs are similar to those in Table 9: a single sequence shows up in the inner table. Further, when a less stringent threshold is chosen, these statistics also identify pairs where a frequent substitution at one position is associated with the consensus amino acid at the other location. Though these statistics are geared towards assessing departures from independence (and so can be regarded as a combination of the binomial and $\chi^2$ tests), they fail to select extremely strong linkages elicited via either the $\chi^2$-test - (13, 25), (18, 25) and (20, 25) - or the binomial test - (11, 13), (32, 34).

TABLE 9
*Data for positions 7 and 30.*

| | | Position 30 | |
|---|---|---|---|
| | | I | V |
| Position 7 | N | 353 | 0 |
| | Q | 1 | 0 |
| | T | 0 | 1 |
| | Y | 1 | 0 |

We recommend that the reference distributions for the statistics described in Section 3 be computed under both scenarios: independence of the sequences and phylogenetic relationships among the sequences. Though the model for the evolutionary process is somewhat simplistic, it most often generates a distribution with a wider range. With enough sequence data available, these tests can be generalized to deal with triplets, quadruplets, ..., of positions.

## REFERENCES


[1] A. AGRESTI, *Categorical Data Analysis*, Wiley, New York, 1990.

[2] P. BICKEL, P. COSMAN, R. OLSHEN, P. SPECTOR, A. RODRIGO, AND J. MULLINS, *Covariability of V3 loop amino acids*, AIDS Research and Human Retroviruses, 12 (1996), pp. 1401–1411.

[3] S. BLOWER AND A. MCLEAN, *Prophylactic vaccines, risk behavior change, and the probability of eradicating HIV in San Francisco*, Science, 265 (1994), pp. 1451–54.

[4] B. EFRON AND R. TIBSHIRANI, *An Introduction to the Bootstrap*, Chapman and Hall, New York, 1993.

[5] J. FELSENSTEIN, *Phylogenies from molecular sequences: Inference and reliability*, Annual Review of Genetics, 22 (1988), pp. 521–65.

[6] L. GOODMAN AND W. KRUSKAL, *Measures of Association for Cross Classifications*, Springer, New York, 1979.

[7] B. KORBER, R. FARBER, D. WOLPERT, AND A. LAPEDES, *Covariation of mutations in the V3 loop of the human immunodeficiency virus type 1 envelope protein: An information-theoretic analysis*, Proceedings of the National Academy of Sciences U.S.A., 90 (1993), pp. 7176–80.

[8] B. KORBER AND G. MYERS, *Signature pattern analysis: A method for assessing viral sequence relatedness*, AIDS Research and Human Retroviruses, 8 (1992), pp. 1549–60.

[9] C. OU, C. CIESIELSKI, G. MYERS, C. BANDEA, C. LUO, B. KORBER, J. MULLINS, G. SCHOCHETMAN, R. BERKELMAN, A. ECONOMOU, J. WITTE, L. FURMAN,

G. Satten, K. MacInnes, J. Curran, H. Jaffe, Laboratory Investigation Group, and Epidemiologic Investigation Group, *Molecular epidemiology of HIV transmission in a dental practice*, Science, 256 (1992), pp. 1165–71.

[10] B. Preston, B. Poiesz, and L. Loeb, *Fidelity of HIV-1 reverse transcriptase*, Science, 242 (1988), pp. 1168–71.

[11] F. Seillier-Moiseiwitsch, B. Margolin, and R. Swanstrom, *Genetic variability of the human immunodeficiency virus: Statistical and biological issues*, Annual Review of Genetics, 28 (1994), pp. 559–96.

[12] D. Swofford and G. Olsen, *Phylogeny reconstruction*, in Molecular Systematics, D. Hillis and C. Moritz, eds., Sinauer, Sunderland, MA, 1990, pp. 411–501.

# BIOLOGICAL AND STATISTICAL STUDIES FOR DISEASES INVOLVING mtDNA MUTATIONS*

FENGZHU SUN†

**Abstract.** Mitochondrial DNA (mtDNA) mutations have been shown to be involved in several rare and complex diseases. This paper first presents a brief review on mitochondrial genetics, heteroplasmic mtDNA transmission and biological studies for diseases related to mtDNA mutations. Then we present a detail review on statistical methods for testing mtDNA mutation involvement in diseases and for estimating their contribution to the disease if mtDNA mutations are involved. Available methods for studying the interaction between nuclear and mtDNA mutations are also discussed. The purpose of this paper is to stimulate research in the statistical studies of mitochondrial diseases.

**Key words.** mitochondrial genetics, heteroplasmic mtDNA transmission, mathematical modeling, statistical power.

**AMS(MOS) subject classifications.** 60K99, 62P10, 92D10, 92D30.

**1. Introduction.** Several diseases have been shown to be related to mitochondrial DNA (mtDNA) mutations. MtDNA plays a vital role in producing ATP via the complex oxidative phosphorylation (OXPHOS) pathway. ATP production is important in the function of brain, skeletal muscle, heart, and other organs. Therefore it is hypothesized that mtDNA mutations might be a common cause of human degenerative diseases and aging (Taylor 1992, Wallace 1992). The study of mtDNA mutation involvement in diseases is a rapidly developing field. On the other hand, statistical studies of testing mtDNA mutation involvement in diseases lag far behind. In this paper, we present a review of biological and statistical studies on diseases related to mtDNA mutations with the intention to stimulate further statistical studies to detect mtDNA mutation involvement and to estimate their contribution to the disease if mtDNA mutations are involved.

**2. Mitochondrial genetics.** Human mtDNA is a 16,569 base pair (bp) closed circular molecule located within the matrix of the double membrane mitochondrion. The complete human mtDNA has been sequenced (Anderson et al. 1981). It encodes 37 genes, including a small (12S) and a large (16S) rRNA, 22 tRNAs, and 13 polypeptides. MtDNA encoded polypeptides are subunits of the respiratory chain which is embedded in the inner membrane of the mitochondrion and consists of about 90 different subunits. About 77 subunits of the respiratory chain are also encoded

*This work is partly supported by NIH FIRST Award DK53392 and a grant from the University Research Council of Emory University. Part of the work was done when I was visiting IMA of the University of Minnesota from July 6 to 11, 1997.

†Department of Genetics, Emory University School of Medicine, Atlanta, GA 30322.

by nucleus-encoded polypeptides. Because of the dual roles of nuclear and mitochondrial encoded proteins in the respiratory chain, mutations in both nuclear and mitochondrial DNA can cause diseases related to the respiratory chain. Bu and Rotter (1991) and Bu et al. (1992) classified mitochondrial related diseases into three main classes. The first class involves mutations only in the mitochondrial genome. Because mtDNA is maternally inherited, this type of diseases exhibits maternal inheritance. The second class involves mutations in both the nuclear genome and the mitochondrial genome. The interaction of the mutations in the nuclear genome and the mitochondrial genome gives the disease phenotype. This class of diseases usually presents excess maternal inheritance. The third class involves mutations only in the nuclear genome, in which the nuclear genome produces defective subunits for mitochondrial function. This class of diseases does not necessarily show excess maternal inheritance and can usually be studied by methods for nuclear diseases. In this paper we mainly consider the first two classes of diseases.

Mitochondrial genetics is quite different from nuclear genetics. MtDNA is predominantly transmitted through oocyte cytoplasm and therefore it is maternally inherited (Case and Wallace 1981, Giles et al. 1980). Although low levels of paternal mtDNA transmission has been reported in interspecific mouse crosses (Kaneda et al. 1995), it is considered rare in human. Mutations accumulate extremely rapidly in mtDNA, about 10-20 times the mutation rate in nuclear genes (Neckelmann et al. 1987, Wallace et al. 1987). The high mutation rate is thought to be resulted from the mtDNA's lack of protective histones, inefficient DNA repair system, and continuous exposure to the mutagenic effects of the oxygen radicals generated by oxidative phosphorylation.

Unlike nuclear genes which are present in diploid in each cell, each human cell contains hundreds of mitochondria and thousands of mtDNAs. If a mutation occurs in some of the mtDNAs in a cell, it creates a mixture of normal and mutant mtDNAs: a state known as heteroplasmy. When heteroplasmic cells divide, the mtDNA genotype undergoes replicative segregation and the proportion of mutant mtDNAs drifts, such that cells tend towards having either all mutant or normal mtDNAs: a state known as homoplasmy. The mechanism of heteroplasmic mtDNA transmission is complicated and is a topic of current research. Several biological studies have been done to understand the transmission of heteroplasmic mtDNAs in model organisms.

Solignac et al. (1984) studied the heteroplasmic mtDNA transmission in a strain of *D. mauritiana* of *Drosophila*. Volz-Lingenhohl et al.(1992) studied heteroplasmic mtDNA transmission for a large-scale deletion in the coding region of *Drosophila subobscura* mtDNA. In both studies, the heteroplasmic mtDNA transmission is very stable in *Drosophila* and no homoplasmy was found after several generations. Rand and Harrison (1986) studied heteroplasmic mtDNA transmission in crickets trying to find the distribution of the fraction of mutant in the offspring from heteroplasmic females. As in *Drosophila*, heteroplasmic mtDNA transmission is also stable in *crickets*. The Wright-Fisher model from population genetics fits the mtDNA transmission data well. Extensive studies have been done on heteroplasmic mtDNA transmissions in bovines and Holstein cows (Hauswirth and Laipis 1982, 1985, Ashley et al. 1989, Laipis et al. 1988, Koehler et al. 1991). In contrast to the studies in *Drosophila* and *crickets*, heteroplasmic mtDNAs can rapidly achieve homoplasmic state in a few generations. They proposed that at some stage in oogenesis the number of mitochondrial genomes within any one developing oocyte is reduced to as few as five or less. This is called the bottleneck model. To understand when the bottleneck occurs, Jenuth et al. (1996) created a heteroplasmic mouse model to study mtDNA distributions in mature and progenitor oocytes. They concluded that random segregation and a bottleneck of about 200 mtDNAs can explain their data. They also concluded that the major changes in the distribution of hteroplasmic mtDNAs occurs during the development from primary oocyte to mature oocyte. The study of heteroplasmic mtDNA transmission in humans is more difficult due to the small number of heteroplasmic individuals available. Rapid changes in mtDNA genotype have been observed in families affected with Leber hereditary optic neuropathy (LHON) before and were taken as evidence that the bottleneck might be very narrow in humans. On the other hand, Howell et al. (1992) studied the segregation of a heteroplasmic silent mtDNA polymorphism in a multigeneration family with a homoplasmic LHON mtDNA mutation and found that heteroplasmy is maintained within this family. They concluded that the bottleneck might not be too narrow. In a more recent study skewed segregation of the mtDNA mutation in human oocytes was reported (Blok et al. 1997). More studies need to be done to understand heteroplasmic mtDNA transmission in humans.

In an individual harboring heteroplasmic mtDNAs, different tissues may have different fractions of mutant mtDNAs. This is called tissue specificity. Even different cells in the same tissue may have different fractions of mutant mtDNAs. Different tissues and organs rely on mitochondrial energy to various extents, with the central nervous system, followed by heart, muscle, kidney, and endovine. As the proportion of mutant mtDNAs increases, mitochondrial energy output declines. When the ATP-generating capacity of the tissue falls below the energy threshold necessary for normal tissue function, disease occurs. As a result, these tissues are more likely to be affected by mitochondrial mutations. In most of the mitochondrial diseases discovered so far, one or more of these tissues are affected. Due to the variability of heteroplasmic mtDNA transmission and tissue specificity, diseases related to mtDNA mutations can show a wide range of variability even within the same family.

**3. Mitochondrial diseases.** MtDNA mutations were first shown to be involved in LHON and mitochondrial myopathies using molecular techniques in 1988 (Holt et al. 1988, Wallace et al. 1988). Three mtDNA mutations: a G-A transition at nucleotide 3460, a G-A transition at nucleotide 11778 and a T-G transition at nucleotide 14484, account for about 70-90% of the worldwide LHON patients. The three mutations have never been found in non-LHON controls and are thus referred as primary mutations. About 15 other mtDNA mutations have been found to be associated with LHON but the pathological role of these mutations are not established yet as they either were found only in a single pedigree or are present in low frequency in controls. These mutations are referred as secondary mutations. Secondary mutations may either be simple polymorphism and their association with LHON reflects historical genetic structure, or they may have pathological role in the development of LHON. There is no consensus about the classification of primary and secondary mutations.

To understand the effect of the three primary mtDNA mutations on LHON under different genetic backgrounds, we recently performed a phylogenetic approach for LHON cases and non-LHON controls (Brown et al. 1997). Extensive molecular studies have shown that each population can be divided into several different lineages according to their evolutionary history. For example, it has been shown that about 65% of North American Caucasians belong to four major lineages: H, I, J, and K. Each lineage is defined by one or two neutral mtDNA mutations which are specific to that lineage. We collapse all the other individuals who do not belong to the above four lineages into one lineage denoted by R. Using 175 controls and 17 11778-, 10 3460-, and 8 14484-positive patients, we found that 3460-positive individuals distribute along the different lineages proportional to that of controls. In contrast, 6 out of the 8 14484-positive individuals belong to lineage J while only 16 out of the 175 controls belong to lineage J. The difference is statistically significant even after adjusting for multiple comparisons (p-value = 0.0001). 11778-positive individuals are also more likely to belong to lineage J than non-LHON controls although the difference is not statistically significant after adjusting for multiple comparisons (p-value = 0.0442). To confirm our observations, we collected all the data on LHON patients from the literature and found that the same conclusions hold for this data set too. The reason(s) for the difference is not clear and needs to be further studied. This approach may prove to be useful to study the pathogenicity of certain mtDNA mutations. In order for this method to work, accurate methods for defining the different lineages are needed.

Although many studies have been done on LHON, its pathogenesis is still not well understood. LHON is strictly maternally inherited and therefore mtDNA mutations must account for most of the LHON cases if not all. On the other hand, males are more likely to be affected than females. It has been found that about 80-85% of LHON patients are males. Thus mtDNA mutations can not be the only factor causing LHON. Nuclear mutations might also be involved in LHON. It is hypothesized that an X-linked factor interacting with mtDNA mutations to give the LHON phenotype. An initial linkage to DXS7 on the X-chromosome was reported (Vilkki et al. 1991) and was later reevaluated and excluded (Juvonen et al. 1993). This locus was also excluded in other families (Sweeney et al. 1992). No mechanisms have been found to explain the predominant prevalence of LHON among males over females.

Besides LHON, several other rare diseases including Kearns-Sayre syndrome (KSS), myoclonic epilepsy and ragged-red fibre disease (MERRF), and mitochondrial encephalomyopathy, lactic acidosis, and stroke like episodes (MELAS) have been shown to be associated with mtDNA mutations.

MtDNA mutations were also found in certain types of diabetes mellitus patients (Ballinger et al. 1992, Reardon et al. 1992, van den Ouweland et al. 1992, Gerbitz et al. 1995). Indications of mtDNA mutation involvement in diabetes mellitus came from two sources. First, epidemiological studies always find excess maternal inheritance for non-insulin dependent diabetes mellitus (NIDDM). Second, diabetes mellitus is frequently associated with mitochondrial diseases. A mutation at nuclear position 3243 was first found in a large diabetes pedigrees in 1992 (van de Ouweland et al. 1992). About 1.5% of diabetes families have been shown to harbor this mutation (Gerbitz et al. 1995). Although mtDNA mutation involvement in diabetes can explain excess maternal inheritance in NIDDM, this small fraction of mtDNA mutation in NIDDM can not explain the strong excess maternal inheritance observed for NIDDM. In a recent study, we estimated that about 22% of NIDDM maybe due to mtDNA mutations with 95% confidence interval 6 to 38% (Sun, unpublished data). Therefore more studies need to be done to estimate the contribution of mtDNA mutations to diabetes mellitus.

**4. Statistical methods for studying mitochondrial diseases.** Compared to molecular studies of mitochondrial diseases, statistical study on mtDNA mutation involvement in diseases is an underdeveloped subject although several investigators previously addressed this problem. In this section, we review statistical studies to test mtDNA mutation involvement in diseases.

**4.1. Heteroplasmic mtDNA transmission.** As discussed above, the mechanism of heteroplasmic mtDNA transmission is not well understood. It is important both in itself theoretically and in studying diseases related to mtDNA mutations. Given a female having a fixed fraction of mutant mtDNAs in her oocyte, what is the distribution for the fraction of mutant mtDNAs in her offspring? This question can be compared with the Mendelian segregation rule for nuclear genes although the later is much simpler and well understood. To understand the mechanism of heteroplasmic mtDNA transmission mathematically, Hopfenmuller (1978, 1979) constructed a mathematical model for heteroplasmic mtDNA transmission in germ cells and somatic cells based on the knowledge of oocyte development. The oocyte development was divided into four stages: the cleavage period, the cell propagation period, the cell growth period, and the cell turn over period.

In the cell cleavage period, it was assumed that a human oocyte containing about 30,000 mitochondria is divided into $2^6$ to $2^8$ cells after 6 to 8 cell divisions. After the cleavage period, each cell contains about 300 mitochondria. The daughter cells were referred as primordial germ cells (PGC) in the biology literature. During this period, each mitochondria belongs to a PGC with equal probability $2^{-i}$, where $i$ is the number of cell divisions in the cleavage period. Let $A_F$ be the number of mutant mitochondria in the original cell. Then the number of mutant mitochondria in a PGC will be binomially distributed, $B(A_F, 2^{-i})$.

In the cell propagation period, it was assumed that each PGC derived from the cell cleavage period undergoes mitotic cell division without changing the total number of mitochondria in the cell. It was assumed that a PGC goes through 17–21 cell mitochondrial cell divisions in this period. The resulting daughter cells are called primary oocytes. Let $n$ be the number of mitochondria in a cell and $\beta$ be the number of mutant mitochondria after the $k$-th cell division. In the $k+1$-st mitotic cell division the cell first double the number of mitochondria to $2n$ mitochondria with $2\beta$ mutant. The model assumes that these $2n$ mitochondria separates into two cells with $n$ mitochondria in each cell. The number of mutant mitochondria in each of the two cells is modeled as a geometric random variable. The probability of having $m$ mutant mitochondria is given by

$$p_{k+1}(m) = \sum_{\beta=m/2} \binom{2\beta}{m}\binom{2n-2\beta}{n-m} p_k(\beta) / \binom{2n}{n}.$$

In the growth period, the primary oocytes with about 300 mitochondria develop into mature oocytes with about 30,000 mitochondria. It was assumed that, during this period, each time a mitochondria is randomly sampled and then divides into two mitochondria until the cell is mature. If the initial cell has $n$ mitochondria, the cell will contain $n+k$ mitochondria after the $k$-th cell division. Let $p_k(m)$ be the probability of having $m$ mutant mitochondria after the $k$-th cell division. Then $p_k(m)$ satisfies the following recursive equation

$$p_{k+1}(m) = \frac{m-1}{n+k} p_k(m-1) + \frac{n+k-m}{n+k} p_k(m).$$

The first term is the probability that there are $m-1$ mutant after the $k$-th cell division and one mutant mitochondria was selected to duplicate. The second term is the probability that there are $m$ mutant after the $k$-th cell division and one wide type mitochondria was selected to duplicate.

In the turn over process, it was assumed that first a mitochondria is randomly sampled and duplicated, and then one of the resulting mitochondria is selected to be lost. Let $p_k(m)$ be the probability of having $m$ mutants after the $k$-th turn over process assuming that there are a total of $n$ mitochondria in the cell. Then

$$p_{k+1}(m) = (1 - \frac{m}{n+1})\Big\{p_k(m-1)\frac{m-1}{n} + p_k(m)(1 - \frac{m}{n})\Big\} + \frac{m+1}{n+1}\Big\{p_k(m)\frac{m}{n} + p_k(m+1)(1 - \frac{m+1}{n})\Big\}.$$

Like any mathematical model, this model can not be completely consistent with the real mechanism underlying heteroplasmic mtDNA transmission. The question is whether this model captures the main features of heteroplasmic mtDNA transmission. It is also too complicated to obtain any practical results. The validity of this model was not compared to real data sets. The model received little attention since its publication. Now data on heteroplamsmic mtDNA transmissions are available on model systems such as mouse (Jenuth et al. 1996). It is interesting to see if this model fits the data or not. From their data set, Jenuth et al. (1996) concluded that the main changes in the fraction of mutant mitochondria occurs from the transition from PGC to primary oocytes, while the above mathematical model predicts that the main changes occur from primary oocytes to mature oocytes. Thus the model for the propagation period may not represent the real mechanism. Jenuth et al. (1996) also concluded that the Wright—Fisher model on random drift can explain their data. If this is true for all model organisms and human, the model for heteroplasmic mtDNA mutations might be much simpler than had been thought. Both biological and mathematical understanding of heteroplasmic mtDNA transmission are needed.

**4.2. Testing for mtDNA mutation involvement in diseases.** Because mtDNA is maternally inherited, diseases related to mtDNA mutations must show excess maternal inheritance. Excess maternal inheritance indicates mtDNA mutation involvement although several other factors also result in excess maternal inheritance. Several investigators studied this problem for various diseases. It was observed long ago that LHON is exclusively maternally inherited leading to the hypothesis that mtDNA mutations might be involved in LHON. It was later confirmed using molecular approaches. Ottman et al. (1988) used standard epidemiologic methods to demonstrate that offspring of mothers with epilepsy are more likely to be affected than those with affected fathers. Based on this observation and other facts, they proposed that mtDNA mutations or intrauterine, neonatal, or early childhood environmental factors might be involved in the etiology of epilepsy. Mili et al. (1996) extended their approach to screen for excess maternal inheritance in extended pedigrees. They compared the risk of the disease among individuals who have affected mother or maternal grandmother or maternal aunts or uncles with the risk of the disease among individuals with affected father or paternal grandparents or paternal aunts or uncles. They applied the approach to a data set on LHON disease and found excess maternal inheritance while they did not find any excess maternal inheritance in a data set on bipolar affective disorder, a psychiatric disorder having a population prevalence of about 1%. MtDNA mutations have been implicated for bipolar affective disorder. Through unilineal family studies of bipolar affective disorder, two groups (McMahon et al. 1995, Gershon et al. 1996) proposed a novel hypothesis that mtDNA mutations might play a role in bipolar affective disorder. The scheme used by McMahon et al. (1995) was to sample probands with at least two affected sibs or one affected sib and one (only one) affected parent. They found more affected mothers than affected fathers of the probands. They also found that maternal relatives are more likely to be affected than non-maternal relatives of the probands. In Gershon et al. (1996), they sampled pedigrees with at least six affected individuals. Similar results were obtained in this data set as that in McMahon et al. (1995).

The above studies did not consider the power of the proposed test under various disease transmission models involving mtDNA mutations. One question is what type of sampling schemes are most powerful in detecting the involvement of mtDNA mutations if it exists. To address this problem, we recently studied the power of the test by comparing the recurrence risk of certain type relatives of probands along the mitochondrial lineage with that of the same type relatives of probands along the non-mitochondrial lineage (Sun et al. 1998). The mitochondrial lineage of a proband is defined as the proband's relatives who share the same mtDNA assuming homoplasmic mtDNA transmissions. The other relatives belong to the non-mitochondrial lineage. In order to study the power of the tests under the hypothesis of mtDNA mutation involvement using current statistical theory and to remove the dependency among family members, we assumed that only one randomly chosen relative of the proband was selected from each family. In that study, we considered a heterogeneity model in the sense that the disease can either be caused by mutations at nuclear loci or in the mitochondrial genome. Under this model, the power of the test increases as the relationship between the probands and their relatives becomes more distant. On the other hand, under a multiplicative epistatic model in which mutations at the nuclear loci and the mitochondrial genome interact with each other to give the disease phenotype, the power of the test decreases as the relationship between the probands and the relatives becomes distant. From the above study we see that the usefulness of different types of proband-relative pairs using the proposed test to detect mtDNA mutation involvement depends on the underlying disease transmission model. To find the best sampling scheme, it is important to distinguish hetrogeneity model from multiplicative model. Risch's method (1990) might be adapted to achieve this goal.

The problem with the above approach is that factors other than mtDNA mutation involvement might also result in excess maternal inheritance such as X-linkage, maternal imprinting with activation of disease alleles transmitted by the mother, recall biases, different prevalences of the disease in males and females, etc.. In Sun et al. (1998) we proposed methods to distinguish mtDNA mutation involvement from these factors using proband-relative pairs.

Shork and Guo (1993) considered testing mtDNA mutation involvement in diseases in the traditional segregation analysis framework. There are three main components in traditional segregation analysis: (i) the penetrance function, $\phi$, which gives the probability of having the disease given an individual's genotype; (ii) the transmission probability, $\tau$, which gives the probability distribution of an individual's genotype given both parents' genotype, and (iii) the allele frequency parameters, $\gamma$. Given the above three components, it is possible to calculate the likelihood of the individuals' phenotypes in a given pedigree. For multiple pedigrees, the log likelihood function is the sum of the log likelihood over all the pedigrees. Elston and Steward (1971) presented efficient algorithms to calculate the log likelihood function for complex pedigrees. Schork and Guo (1993) provided a variety of pedigree models for diseases involving mtDNA mutations. In the simple maternal inheritance model, the transmission probability function is defined by

$$\tau(g_i = 1 \Big| g_m, g_f) = \begin{cases} 1 & \text{if } g_m = 1 \\ \psi & \text{if } g_m = 0 \end{cases}$$

where $g_i, g_m$ and $g_f$ denote the mitochondrial genotypes of the individual, mother and father, respectively and "1" and "0" denote that the individual has the mutant and non-mutant mitochondrial, respectively. $\psi$ is usually set to be very small and it represents mutation rate at the mitochondrial locus. The penetrance function is defined by

$$\phi(d_i = \text{affected} \Big| g_i) = \begin{cases} \rho & \text{if } g_i = 1 \\ \hat{\phi} & \text{if } g_i = 0 \end{cases}$$

where $\hat{\phi}$ represents the phenocopy probability. For heteroplasmic mtDNA transmission, they proposed to divide the unit interval into several subintervals. The transmission probability is modeled as the probability that the fraction of mutant in an offspring is in interval $j$ given the fraction of mutant in the mother is in interval $i$. This will introduce too many parameters into the likelihood function. Schork and Guo also proposed models for threshold effect and for interaction between nuclear and mtDNA mutations. The problem with these approaches is that a model needs to be specified for the analysis. If the model is correct, the proposed method is supposed to be most powerful in detecting the mtDNA mutation involvement. But in general the correct model is rarely known as a prior.

Once mtDNA mutations have been shown to be involved in the etiology of a disease, it is important to know if mtDNA mutations are the primary cause of the disease or nuclear mutations might also be involved. Bu and Rotter (1991) and Bu et al. (1992) addressed this problem in their studies of LHON. As shown above that mtDNA mutations alone can not explain the LHON transmission data as males are more likely to be affected than females. They hypothesized that an X-linked mutation together with X-chromosome inactivation interacts with mtDNA mutations to give the LHON phenotype. To test this hypothesis, they considered individuals in the pedigrees who are genetically related through females which they called maternal line pedigrees. Because all the members in the maternal line pedigrees have the same mtDNA mutation, the study of the segregation pattern along the nuclear locus is reduced to the one locus model.

Some other diseases have long been shown to be related to nuclear mutations, but nuclear mutations alone can not explain the transmission pattern either. For example, Huntington disease is a classic example of autosomal dominant disease. It has been observed that offspring of affected females have a later age of onset that those of affected males which can not be explained by nuclear mutations alone. Several models have been proposed to explain this observation. Among these models, Boehnke et al. (1983) found that a model assuming a protective factor in the mitochondrial genome combined with the autosomal mutation delays the age of onset for Huntington disease is consistent with the transmission data. This model has not been confirmed by molecular studies.

**5. Discussion.** Identifying genes affecting complex diseases, such as cancer, diabetes, hypertension, and affective disorders, is an important topic in current genetic research. Biological studies have identified over 60 mtDNA point mutations and hundreds of mtDNA rearrangements associated with human diseases (Wallace 1994, 1995). Heteroplasmic mtDNA transmission is a complicated process and is not well understood. Due to heteroplasmic mtDNA transmission, the fraction of mutant mtDNAs can change over generations along matrilineal lineages. In a heteroplasmic individual, different tissues can have different fraction of mutant mtDNAs. Also different tissues have different needs for mitochondrial energy. Thus diseases related to mtDNA mutations can show complex transmission patterns. In studying complex diseases, not only do we need to consider nuclear complexities, such as reduced penetrance, phenocopies, polygenic effect, and heterogeneities, we also need to consider the possible involvement of mtDNA mutations. Failure to take mtDNA mutations into account may give misleading results.

In this brief review we present an outline of biological and statistical studies for diseases related to mtDNA mutations with an intention to stimulate research in developing statistical methods to test mtDNA mutation involvement and to estimate the contribution of mtDNA mutations in the etiology of a disease. Several problems have not been addressed in the literature and need to be studied. First, for common diseases, it is more likely that both nuclear and mtDNA mutations are involved in the diseases. Then it is important to know how nuclear and mtDNA mutations interact with each other to give the disease phenotype: whether either mutation is enough to cause the disease or both mutations are needed to show the phenotype. This study is important for clinical purposes too. For example, a subset of diabetes related to mtDNA mutations have some distinct phenotypic features not showing in other types of diabetes and may represent a heterogeneity model. LHON represents another example that both nuclear and mtDNA mutations are needed to show the phenotype although the responsible nuclear mutation(s) has not been identified yet. Second, once mtDNA mutations have been identified to be involved in the disease, how to estimate the attributable fraction due to mtDNA mutations? Third, what effects can mtDNA mutation involvement have in our effort in identifying nuclear genes affecting the disease using current linkage analysis method such as pedigree method, sib-pair analysis and transmission disequilibrium test (TDT)? If these methods are not powerful enough, are there any better method to find linkage to nuclear genes in the presence of mtDNA mutation involvement? Due to the difficulties in dealing with gene-gene interaction in general, these problems are challenging.

## REFERENCES


[1] Anderson S., Bankier A.T., Barrell B.G., de Bruijn MHL et al., Sequence and organization of the human mitochondrial genome. Nature 9:457–465 (1981).

[2] Ashley M.V., Laipis P.J., Hauswirth W.W., Rapid segregation of heteroplasmic bovine mitochondria. Nucl. Acids Res. 18:7325–7331 (1989).

[3] Ballinger S.W., Shoffner J.M., Hedaya E.V., et al., Maternally transmitted diabetes and deafness associated with a 10.4 kb mitochondrial DNA deletion. Nature Genetics 1:11–15 (1992).

[4] Blok R.B., Gook D.A., Thorburn D.R., Dahl H.M., Skewed segregation of the mtDNS nt 8993 (T→G) mutation in human oocytes. Am. J. Hum. Genet. 60:1495–1501 (1997).

[5] Boehnke M., Conneally P.M., Lange K., Two models for a maternal factor in the inheritance of Huntington disease. Am. J. Hum. Genet. 35:845–860 (1983).

[6] Brown M.D., Sun F.Z., Wallace D.C., Clustering of Leber's hereditary optic neuropathy patients on a Caucasian mitochondrial lineage. Am. J. Hum. Genet. 60:381–387 (1997).

[7] Bu X., Rotter J.I., X chromosome-linked and mitochondrial gene control of Leber hereditary optic neuropathy: Evidence from segregation analysis for dependence on X chromosome inactivation. Proc. Natl. Acad. Sci. USA 88:8198–8202 (1991).

[8] Bu X., Yang H.Y., Shohat M., Rotter J.I., Two locus mitochondrial and nuclear gene models for mitochondrial disorders. Genet. Epidemiol. 9:27–44 (1992).

[9] Case J.T., Wallace D.C., Maternal inheritance of mitochondrial DNA polymorphisms in cultured human fibroblasts. Somat. Cell Genet. 7:103–108 (1981).

[10] Giles R.E., Blanc H., Cann H.M., Wallace D.C., Maternal inheritance of human mitochondrial DNA. Proc. Natl. Acad. Sci. USA 78:5768–5772 (1980).

[11] Gerbitz K.D., van den Ouweland J.M.W., Maassen J.A., Jaksch M., Mitochondrial diabetes mellitus: a review. Biochimica et Biophysica Acta 1271:253–260 (1995).

[12] Gershon E.S., Badner J.A., Detera-Wadleigh S.D., Ferraro T.N., Berrettini W.H., Maternal inheritance and chromosome 18 allele sharing in unilineal bipolar illness pedigrees. Am. J. Med. Genet. 67:202–207 (1996).

[13] Hauswirth W.W., Laipis P.J., Mitochondrial DNA polymorphism in a maternal lineage of Holstein cows. Proc. Natl. Acad. Sci. USA 79:4686–4690 (1982).

[14] Hauswirth W.W., Laipis P.J., Transmission genetics of mammalian mitochondria: A molecular model and experimental evidence. in *Achievements and perspectives in Mitochondrial Research*, ed. Quagliarello, Z. (Elsevier, Amsterdam), 2:49–59 (1985).

[15] Holt I.J., Harding A.E., Morgan Hughes J.A., Deletions of muscle mitochondrial DNA in patients with mitochondrial myopathies. Nature 331:717–719 (1988).

[16] Hopfenmuller W., A mathematical model for extrachromosomal heredity. Biometrical J. 20:609–618 (1978).

[17] Hopfenmuller W., A mathematical model for extrachromosomal heredity in somatic cells, exemplified by liver cells. Biometrical J. 21:431–437 (1979).

[18] Howell N., Halvorson S., Kubacka I., McCullough D.A., Bindoff L.A., Turnbull D.M., Mitochondrial gene segregation in mammals: is the bottleneck always narrow? Hum. Genet. 14:146–151 (1992).

[19] Jenuth J.P., Peterson A.C., Fu K., Shoubridge E.A., Random genetic drift in the female germline explains the rapid segregation of mammalian mitochondrial DNA. Nature Genet. 14:146–151 (1996).

[20] Juvonen V., Vilkki J., Aula P., Nikoskelainen E., Savontaus M.L., Reevaluation of the linkage of an optic atrophy susceptibility gene to X-chromosome markers in Finnish families with Leber hereditary optic neuroretinopathy (LHON). Am. J. Hum. Genet. 53:289–292 (1993).

[21] Kaneda H., Hayashi J.I., Takahama S., Taya C., Fisher-Lindahl K., Yonekawa H., Elimination of paternal mitochondrial DNA in intraspecific crosses during early mouse embrogenesis. Proc. Natl. Acad. Sci. USA 92:4542–4546 (1995).

[22] Koehler C.M., Lindberg G.L., Brown D.R., Beitz D.C., Freeman A.E., Mayfield J.E., Myers A.M., Replacement of bovine mitochondrial DNA by a sequence variant within one generation. Genetics 129:247–255 (1991).

[23] Laipis P.J., Van de Walle M.J., Hauswirth W.W., Unequal partitioning of bovine mitochondrial genotypes among siblings. Proc. Natl. Acad. Sci. USA 85:8107–8110 (1988).

[24] McMahon F.J., Stine O.C., Meyers D.A., Simpson S.G., and Depaulo J.R., Patterns of maternal transmission in bipolar affective disorder. Am. J. Hum. Genet. 56:1277–1286 (1995).

[25] Mili F., Flanders W.D., Sherman S.L., Go R.C.P., Wallace D.C., Genetic epidemiology methods to screen for matrilineal inheritance in mitochondrial disorders. Genet. Epidemiol. 13:605–614 (1996).

[26] Neckelmann S.N., Li K., Wade R.P., Shuster R., Wallace D.C., cDNA sequence of a human skeletal muscal ADP/ATP translocator: lack of a leader peptide, divergence from a fibroblast DNA genes. Proc. Natl. Acad. Sci. USA 84:7580–7584 (1987).

[27] Ottman R., Annegers J.F., Hauser W.A., Kurland L.T., Higher risk of seizures in offspring of mothers than of fathers with epilepsy. Am. J. Hum. Genet. 43:257–264 (1988).

[28] Rand D.M., Harrison R.G., Mitochondrial DNA transmission genetics in crickets. Genetics 114:955–970 (1986).

[29] Reardon W., Ross R.J.M., Sweeney M.G., et al., Diabetes mellitus associated with a pathogenic point mutation in mitochondrial DNA. Lancet 340:1376–1379 (1992).

[30] Risch N., Linkage strategies for genetically complex traits. I. Multilocus models. Am. J. Hum. Genet. 46:222–228 (1990).

[31] Schork N.J., Guo S.W., Pedigree models for complex human traits involving the mitochondrial genome. Am. J. Hum. Genet. 53:1320–1337 (1993).

[32] Solignac M., Génermont J., Monnerot M., Mounolou J., Genetics of mitochondria in *Drosophila*: mtDNA inheritance in heteroplasmic strains of *D. mauritiana*. Mol. Gen. Genet. 197:183–188 (1984).

[33] Sun F.Z., Ashley A.E., Durham L.K., Feingold E., Halloran M.E., Manatunga A.K., Sherman S.L., Testing for contributions of mitochondrial DNA mutations to complex diseases. To appear in Genet. Epidemiol (1998).

[34] Sweeney M.G., Davis M.B., Lashwood A., Brockington M., Toscano A., Harding A.E., Evidence against an X-linked locus close to DXS7 determining visual loss susceptibility in British and Italian families with Leber hereditary optic neuropathy. Am. J. Hum. Genet. 51:741–748 (1992).

[35] Taylor R., Mitochondrial DNA may hold a key to human degenerative diseases. J. NIH Res. 4:62–66 (1992).

[36] van den Ouweland J.M.W., Lemkes H.H.P.J., Ruitenbeek W,. et al., Mutation in mitochondrial tRNA$^{Leu(UUR)}$ gene in a large pedigree with maternally transmitted type II diabetes mellitus and deafness. Nature Genetics 1:368–371 (1992).

[37] Vilkki J., Savontaus M.L., Aula P., Nikoskelainen E.K., Optic atrophy in Leber hereditary optic neuroretinopathy is probably determined by an X-chromosome gene closely linked to DXS7. Am. J. Hum. Genet. 48:486–491 (1991).

[38] Volz-Lingenhohl A., Solignac M., and Sperlich D., Stable heteroplasmy for a large-scale deletion in the coding region of *Drosophila subobscura* mitochondrial DNA. Proc. Natl. Acad. Sc. USA 89:11528–11532 (1992).

[39] Wallace D.C., Mitochondrial genetics: a paradigm for aging and degenerative disease? Science 256:628–632 (1992).

[40] Wallace D.C., Mitochondrial DNA sequence variation in human evolution and disease. Proc. Natl. Acad. Sci. USA 91:8739–8746 (1994).

[41] Wallace D.C., Mitochondrial DNA variation in human evolution, degenerative disease, and aging. Am. J. Hum. Genet. 57:201–223 (1995).

[42] Wallace D.C., Singh G., Lott M.T., Hodge J.A., Schurr T.G., Lezza A.M.S., Elsas, L.J., Nikoskelainen E.K., Mitochondrial DNA mutation associated with Leber's Hereditary Optic Neuropathy. Science 242:1427–1430 (1988).

[43] Wallace D.C., Ye J.H., Neckelmann S.N., Singh G., Webster K.A., Greenberg B.D., Sequence analysis of cDNAs for the human and bovine ATP synthase b-subunit: mitochondrial DNA genes sustain seventeen times more mutations. Curr. Genet. 12:81–90 (1987).

# STATISTICAL METHODS IN HUMAN GENETICS

W.J. EWENS*

**1. Introduction.** Statistics and human genetics are twin subjects, having grown with the century together, and there are many connections between the two. Some fundamental ideas in statistics, in particular the concept of the Analysis of Variance, first arose in human genetics, while statistical and probabilistic methods are now central to many aspects of the analysis of questions in the human genetics.

The aim of this paper is to describe three particular applications of statistical methods in human genetics. The first two of these relate to the Human Genome Project, while the third concerns linkage analysis, especially linkage analysis as used for finding disease genes. Many other areas could be discussed, and the three described below were chosen to give an indication of the variety of probabilistic and statistical methods used in human genetics, as well as reflecting the interests of the author.

**2. Reconstructing the genome – clones and anchors.** For the Human Genome Project, and similar projects for other species, we can discuss (i) how can one reconstruct a chromosome's DNA sequence, and (ii) what analyses can one make of it once it is built up? We consider the first question in this section and the second in the following section.

A technical problem with the laboratory manipulation of genetical material is that only short segments of DNA can be handled at any one time. Thus while we might have found the DNA sequence in some small piece of DNA, we are still left with the problem of building up the entire chromosome, or at least as much of it as we can, from the segments available to us.

One way of building up the DNA sequence of a chromosome is by a method using clones and entities generically called anchors. In practice a clone is a length of DNA such as a YAC (yeast artificial chromosome), described for example by Schlessinger (1990). An anchor is thought of as being essentially a point, and in practice might be a restriction site (Botstein et al., 1980), primed PCR products (Williams et al., 1990) or sequence-tagged sites (Olson et al., 1989). Clones and anchors are used together to form contigs, or "anchored islands", and, leaving aside the (very complex) biology, the genome is put together, at least in part, using these anchored islands, (which, for brevity, we henceforth call simply "islands").

Abstracting to a purely mathematical description of the procedure, clones can be thought of respectively as sticks (of fixed or variable size) and anchors as points, or staples. We suppose we have $M$ such sticks

*Department of Biology, University of Pennsylvania, Philadelphia, PA 19104-6018; wewens@sas.upenn.edu.

(clones), $N$ staples (anchors), and are considering a segment of the real line (the chromosome) of length $G(G >> L)$. For the moment, we assume that all clones have the same length $L$.

In the simplest case, the clones and anchors can be thought of as having been randomly thrown on a line segment of length $G$. By "randomly" we mean according to two independent homogeneous Poisson processes. An anchored island is a collection of one or more clones "stapled together", and also stapled to the chromosome, by one or more anchors. Unstapled clones, that is clones which have no anchor falling on them, are not useful biologically, and islands are constructed only from stapled, that is anchored, clones.

What we know of the genome is what we know about these anchored islands, specifically their location, DNA content, and size. The proportion of the chromosome whose DNA sequence we can establish by this method is the proportion of the total length $G$ of the chromosome covered by anchored islands. From the biological point of view, it is then natural to ask, given $N, M, L$ and $G$: (i) What is the mean proportion of the genome covered by (anchored) islands? (ii) What is the mean number of islands formed by this anchoring process? There are other questions we can ask of biological relevance, such as "what is the mean island size?", but we concentrate here on the above two.

Note that, if anchors can be assumed to be everywhere dense, (so that all clones are stapled, and all form islands or parts of islands), there are parallels with queuing theory. If we imagine the real axis to be time (rather than the chromosome), that customers arrive at random times, to be served by infinitely many servers, with constant time $L$ for any customer, then a busy time in the queueing system corresponds to a portion of the genome covered by an island. The mean length of a busy period is the mean island size and the mean proportion of the time that one or more customers are being served corresponds to the mean proportion of the genome covered by islands. But the genetic process is more complex than the queuing process, because of the anchoring complication.

We outline here the calculation of (i) the mean number of islands, and (ii) the mean proportion of the chromosome covered by islands. For both calculations it is convenient to introduce the dimensionless quantities $a$ and $b$, defined by

$$a = LN/G, \quad b = LM/G, \tag{1}$$

as well as the quantities $\alpha$ and $\beta$, defined by

$$\alpha = N/G, \quad \beta = M/G. \tag{2}$$

These parameters have the following interpretation. $\alpha$ and $\beta$ are the Poisson rates for clones and anchors, respectively: the probability that the left-hand end of a clone occurs in $(x, x + \delta x)$ is $\alpha\, \delta x$, and the probability that an anchor arises in the same interval is $\beta\ \delta x$. The parameter $a$ is sometimes called the coverage: it is the ratio of the sum of all clone lengths divided by the length of the chromosome. Equivalently, it is the mean number of clones to cross any given point, and since this number has a Poisson distribution, the probability that a randomly chosen point is not covered by a clone is $e^{-a}$. Similarly, $b$ is the mean number of anchors on any given clone.

From these we can calculate other important quantities. For any point $P$ the length (to the right of $P$) of any clone crossing $P$ has a uniform distribution on $(0, L)$. If $k (k \geq 1)$ clones cross the point $P$, there will be some "rightmost" of these clones. The segment length $x$ to the right of $P$ of this clone may be called the "rightmost projection" of the clones covering $P$. The density function of this rightmost projection is, from standard order statistics theory,

$$f(x) = kx^{k-1}/L^k. \tag{3}$$

The joint probability that at least one clone crosses $P$ and that this rightmost projection takes the value $x$ is then $g(x)$, where

$$\begin{aligned} g(x) &= \sum_{k \geq 1} \{e^{-a}a^k/k!\} kx^{k-1}/L^k \\ &= aL^{-1}e^{-a+ax/L}. \end{aligned} \tag{4}$$

Similarly the probability that the point $P$ is covered by at least two clones, and that the span of these clones is $z$, can be shown to be $h(z)$, given by

$$h(z) = (2L - z)a^2L^{-2}ei^{-2a+az/L}. \tag{5}$$

(We have been somewhat casual in the use of differential elements in these two statements.)

The mean number of islands can be found in two ways. First, it is the mean number of clones which are the rightmost clones of an island. This is the number of clones ($N$) multiplied by the probability that the clone is anchored, but has no other anchored clone to its right in the same island. This latter requirement is the requirement that the right-hand end of the clone (which we think of as the point $P$ above) is not crossed by any other clone sharing an anchor with the clone in question. From this we find that the mean number of islands is

$$Ne^{-a}(1 - e^{-b}) + N \int_0^L g(x)e^{-\beta x}(1 - e^{-\beta(L-x)})dx,$$

and this reduces to

$$Nb(e^{-a} - e^{-b})/(b - a). \tag{6}$$

The same calculation can be done even more easily by considering anchors rather than clones. The mean number of islands is the mean number of right-most anchors on islands. Thinking of an anchor as the point $P$, this is the number of anchors ($M$) multiplied by the probability that the point $P$ is covered by at least one clone and that there is no further anchor to the right of $P$ on the rightmost projection of the clones covering $P$. This is

$$M \int_0^L g(x) e^{-\beta x} dx,$$

and this reduces to (6).

So far as the mean proportion of the chromosome covered by islands is concerned, it is easier to find the complementary quantity, namely the mean proportion of the chromosome not covered by islands. This is the probability that a randomly chosen point $P$ is not covered by an island, and this probability can be calculated as the sum of three terms. The first term is the probability that the point $P$ is not covered by any clone, namely $e^{-a}$. The second term is the probability that the point $P$ is covered by exactly one clone which, however, has no anchor on it, namely $ae^{-a-b}$. The third term is the probability that the point $P$ is covered by two or more clones with, however, no anchor on the span of these clones. This latter probability is

$$\int_L^{2L} h(z) e^{-\beta z} dz.$$

Evaluation of this using (5) shows that the mean proportion of the chromosome covered by islands is

$$(7) \qquad 1 - e^{-a} - a(b^2 - ab - a)e^{-a-b}/(b-a)^2 - a^2 e^{-2b}/(b-a)^2.$$

Numerical calculations using these formulae appropriate to the plant *Arabodopsis Thaliana* are given by Ewens et al. (1991). For the data available for this plant, $N$ = number of clones = 2300, $M$ = number of anchors = 500, $L$ = length of each clone = 250, $G$ = length of chromosome = 100,000. For these data the mean number of islands is 181 and the mean proportion of the chromosome covered by islands is 0.8657.

In reality, of course, the clone and anchor process not so simple as described above. Different clones have different lengths. The clone and anchor placement processes need not be Poisson, and even if they are, need not be homogeneous. A further complication is that of "hotspotting": the genome may consist of regions where clones and/or anchors are likely to arise, interspersed with regions where they are less likely to arise. Important progress on generalizing the formulae given above to these more realistic situations has been made by Arratia et al. (1991) and by Schbath (1997).

**3. Analyzing the genome - the BLAST algorithm.** For a mathematician or statistician planning to become involved in the quantitative aspects the human genome project, it would be advisable to focus attention on questions that will arise once the genome has been put together, since this will have been done in the next four or five years. The sort of question which one might ask, when the "standard" human DNA sequence (and that of other species) is available, is typified by the following.

Suppose we are given two aligned DNA sequences, such as those in (8), coming perhaps from the genomes of two different species. Do these two sequences appear to be homologous? That is, are they perhaps derived from the same evolutionary ancestor?

$$
\begin{array}{l}
g\,g\,a\,t\,a\,g\,c\,t\,g\,t\,a\,g\,a\,t\,a\,g\,c\,t\,a\,a\,t\,g\,c\,t\,a\,g\,a \\
c\,a\,a\,t\,a\,c\,c\,c\,c\,g\,t\,g\,t\,t\,g\,c\,g\,a\,g\,a\,c\,c\,t\,t\,a\,g\,c
\end{array} \tag{8}
$$

There is a match at some positions $(3, 4, 5, 7, ...)$, but a mismatch at many others $(1, 2, 6, 8, 9, 10, ...)$. How do we test for a significantly good match? More concretely, how do we test the null hypothesis that there is no significant homology between the two DNA sequences, against the alternative hypothesis that there is significant homology?

The statistical part of the celebrated basic local alignment search tool (BLAST) procedure (Karlin and Altschul, 1990, 1993, Karlin and Dembo, 1992) is a beautiful and powerful example of the application of probability theory and statistics which answers questions of this kind, and indeed many more complicated questions. The probability theory aspects of the procedure are largely due to Karlin, and comprises aspects of random walk theory, of renewal theory and asymptotic distribution theory. We do not give the details of these aspects of the theory here, but present only some central results. We introduce the concepts of the BLAST theory by considering two simple cases, and then later discuss BLAST more generally.

*Case* (i) Both sequences have length $n$, no realignment allowed, perfect matching required. Suppose that, if two nucleotides are taken at random, the probability that they are of the same type (i.e. both $a$, both $c$, both $g$ or both $t$) is some given number $p$, calculated from the overall frequencies of these nucleotides. We wish to test the null hypothesis that the two nucleotides at a matching pair of sites on the two chromosomes being compared are no more likely to be of the same type than are two randomly chosen nucleotides. That is, the null hypothesis claims that the probability that two nucleotides at matching sites in the two sequences are of the same type is $p$. The alternative hypothesis claims that there is homology between the two sequences, that is that this probability exceeds $p$.

We consider first at test statistic the length $L$ of the longest *exactly* matching nucleotide segment between the two sequences. To assess the significance of any observed value $L$, it is necessary to find the null hypothesis distribution of $L$, particularly the right-hand tail part of this distribution.

It is convenient to think of matching segments as possibly having length zero. That is, we imagine that, as we proceed along the matched pair of sequences from left to right, we have just reached a non-matching pair of nucleotides. There is now the potential for a matching segment to begin immediately following this mismatch, although this potentially segment will be of length zero if there is a mismatch at the next following pair of nucleotides. With this definition the length $X$ of a given matching segment, or perhaps more accurately "potentially matching" segment, has a geometric distribution with

$$\text{Prob}(X = i) = qp^i, \qquad (i = 0, 1, 2, ...) \tag{9}$$

$$\text{Prob}(X \le y) = 1 - p^{y+1}, \qquad (y = 0, 1, 2, 3, ...) \tag{10}$$

where $q = 1 - p$. Since the null hypothesis probability of a mismatch at any site is $q$, there will be approximately $qn$ potentially matching such segments. The probability that the largest of $qn$ random variables having the distribution (10) is less than or equal to any prescribed value $y$ is, immediately,

$$\{1 - p^{y+1}\}^{qn}. \tag{11}$$

If $y$ and $n$ are both large, this is approximately $\exp\{-qnp^{y+1}\}$, and for purposes of comparison with BLAST formulae, and recalling the definition of $L$, we will write this in the unusual form

$$\text{Prob}(L \le y) \approx \exp[-pq\ n\ \exp(-y \log(1/p)]. \tag{12}$$

*Case* (ii). Both sequences have length $n$, no realignment allowed, perfect matching not required. It is too much to expect perfect matching between two sequences in (say) two different species, even if they are derived from the same ancestor: evolutionary degradation implies that perfect matching will seldom occur. So we now consider well-matching, rather than perfectly matching, segments. We do this by using scores rather than lengths, allocating some positive score for a match at any site and some negative score, i.e. a penalty, for a mismatch. Then we consider aggregate scores for segments of the matched DNA sequences, and the test statistic we use is the aggregate score of the highest scoring segment, rather than the length of the longest "perfect match" segment.

Suppose we score $+1$ at each matching site and $-1$ (i.e. a penalty of 1) at any mismatching site. A more realistic scoring system is described below; this simple $(+1, -1)$ system is used here to describe the main points at issue in the simplest possible way. The essence of the method is to note that we proceed along the two sequences, say from left to right, checking at each site whether we have a match or a mismatch, the accumulated score $S_n$ after $n$ sites have been compared performs a random walk. We assume that $p < 1/2$, so that the null hypothesis mean score $p - q$ at any one site is negative. This implies that we go along two indefinitely long sequences from left to right, the accumulated score eventually drifts down to $-\infty$ when the null hypothesis is true.

This negative drift implies that the accumulated score eventually going through the sequence of (increasingly negative) ladder points at $-1, -2, -3$, ... (A ladder point (Feller, 1957) occurs at the $n$th site if $S_n < S_1, S_n < S_2, ..., S_n < S_{n-1}$). Suppose that there is a ladder point at site $i$ and we define $x_i$ as $x_i = \max_{k \geq i} (S_k - S_i)$. The quantity $x_i$ represents the maximum height achieved by the "non-negative excursion" starting at the ladder point at $(i, S_i)$. This height will be 0 if there is a mismatch at site $i+1$, so that a new ladder point at $(i+1, S_{i+1})$ is immediately reached. We can find the null hypothesis distribution of $x_i$ by standard random walk methods: specifically,

$$\text{Prob}(x_i = x) = \{(q-p)/q\}(p/q)^x, x = 0, 1, 2, 3, ... \tag{13}$$

If $x_i$ is large, there is substantial homology between the two sequences immediately following site $i$. The test statistic we use is $L$, defined by $L = \max x_i$, the maximum positive excursion after a ladder point.

To assess whether an observed value of $L$ is significant, we need to find the null hypothesis distribution of $L$. This requires us to find, first, the mean number of ladder points in a total matched sequence length of $n$, since $L$ is effectively the maximum of this number of $x_i$ values. This mean number is found by first finding the mean distance between consecutive ladder points. The mean distance, written in (14) below as $E(m)$, is found by using the classic sequential analysis formula

$$E(X_1 + ... + X_m) = E(X)E(m). \tag{14}$$

Here $m$ is a random stopping time, specifically the number of steps (or nucleotide sites) in the random walk between one ladder point and the next, and $X_1, ..., X_m$ are the respective sizes of these steps. In the present case $X_1 + ... + X_m \equiv -1$, since the accumulated score decreases by 1 between one ladder point and the next. Further, $E(X_i) = p - q$. Thus $E(m) = 1/(q-p)$. This implies that the mean number of ladder points in a DNA sequence pair of length $n$ is $(q-p)n$. Thus for all practical purposes, $L$ can be taken as the maximum of $(q-p)n$ random variables, each having the distribution given in (13). Elementary calculations show that

$$\text{Prob}(L \leq y) \approx \exp[-(q-p)^2 p/q^2 n(q/p)^{-y}]. \tag{15}$$

Both (12) and (15) can be written in a slightly different form. The moment-generating function of the step size in the random walk being considered is

$$pe^{\theta} + qe^{-\theta}, \tag{16}$$

and the non-zero value of $\theta$ for which

$$pe^{\theta} + qe^{-\theta} = 1 \tag{17}$$

is

$$\theta = \theta^* = \log(q/p). \tag{18}$$

In terms of $\theta^*$, (15) can be rewritten as

$$\text{Prob}(L \leq y) \approx \exp[-(q-p)^2 p/q^2 n(\exp -y\theta^*)]. \tag{19}$$

We can think of the previous example, where exact matching segments were required, as being the particular case of a random walk with a penalty of $-\infty$ for a mismatch. In other words, the step sizes in the walk are either $+1$ or $-\infty$. The analogue of (17) for this walk becomes $pe^{\theta} = 1$, giving $\theta^* = \log(1/p)$. In terms of this value, (12) can be rewritten as

$$\text{Prob}(L \leq y) \approx \exp[-pq\ n\ \exp(-y\theta^*)], \tag{20}$$

and in this form the similarity with (19) is evident.

These two formulae are specific cases of the general BLAST formula, to which we now turn. In doing so we skip over a complication to the theory, namely that in the general BLAST case where scores are "lattice", as above, there is no limiting distribution for $L$. BLAST theory accounts for this fact and gives an upper and a lower bound for the probability that we want. We in effect compute here a close approximation (also given in the general BLAST theory as one of the bounds) for the distribution of $L$.

More generally, BLAST allows a far more general scoring scheme than that described above. In the comparison between two amino acid sequences, for example, a scoring matrix might be used, where a positive score might arise not only for two identical amino acids but also for two different but similar amino acids, while a negative score would arise for two dissimilar amino acids. For nucleotide sequences we can give different scores for, say, two identical nucleotides, an AG (purine) pair, an AC (purine pyrimidine) pair, and so on.

As the simplest generalization of the case discussed above, suppose that we allocate a (positive) score $S$ at any point where the nucleotides in two sequences are the same, and a (negative) score $T$ at any point where the nucleotides in the two sequences differ. As above we consider the null hypothesis case, that is we assume that the probability that the two nucleotides at matching sites are of the same type is $p$, with $q = 1 - p$. Then the mean score at any nucleotide site is, under the null hypothesis,

$$E(X) = pS + qT, \tag{21}$$

and we require this to be negative.

The first part of the BLAST procedure is to find the non-zero solution $\theta^*$ of the moment-generating function equation

$$pe^{s\theta} + qe^{T\theta} = 1, \tag{22}$$

the generalization of the simple random walk equation (17). Unlike the simple random walk case (which admits the explicit solution (18)), this equation cannot be solved explicitly in general. Nevertheless it can always be solved numerically to any required accuracy.

Next, in the general BLAST score system, the successive ladder points of the accumulated score achieved in any particular concrete case are no longer necessarily $-1, -2, -3, \ldots$ Thus the difference in the accumulated score between two successive ladder points is a random variable, denoted $Z^-$. It is possible to find the distribution of $Z^-$ using a "decomposition of paths" method, which we do not describe here - see Karlin and Dembo (1992) for details. From the distribution of $Z^-$ one can calculate the value $E(Z^-)$, and from this one calculates $A$, the mean distance between successive ladder points using (14), which in this case becomes

$$E(Z^-) = (pS + qT)A, \tag{23}$$

the first term on the right-hand side coming from (21).

With the various calculations just made in hand, we next compute the quantity $K^*$, defined by

$$K^* = [1 - E\{\exp(\theta^* Z^-)\}]^2 / [\{\exp\theta^* - 1\}A^2 E\{X\exp(\theta^* X)\}]. \tag{24}$$

Subject to the approximation described above, The BLAST theory then asserts that if $L$ is the maximum of the various "positive excursion" scores starting at the various ladder points, then

$$\text{Prob}(L \le y) = \exp\{-K^* n \exp(-y\theta^*)\}. \tag{25}$$

This is a central formula of BLAST theory, since is used to assess whether any observed value of $L$ is significant.

This formula is a generalization of both (19) and (20), and we check this by deriving (19) and (20) from it. In the simple random walk case corresponding to (19),

$$Z^- \equiv -1, A = (q-p)^{-1}, \theta^* = \log(q/p).$$

Further,

$$E\{X\exp(\theta^* X)\} = -qe^{-\theta^*} + pe^{\theta^*} = q - p,$$

and from this

$$K^* = \{1 - e^{-\theta^*}\}^2 / [\{e^{\theta} - 1\}\{(q-p)^{-2}\}\{q-p\}] = (q-p)^2 p/q^2.$$

Using this value, (25) is easily seen to reduce to the (19).

If we require two subsequences to match exactly, the penalty $T$ for a mismatch is $-\infty$, so that $Z^- = -\infty$. With this definition of $Z^-$, the term $E\{\exp(\theta^* Z^-)\}$ in the definition of $K^*$ is zero. Further, the expression $E\{X\exp(\theta^* X)\}$ becomes $p\exp\theta^* + 0$, and from the value $\theta^* = \log(1/p)$ applying for this case, this expression reduces to 1. The mean distance $A$ between the starting points of consecutive matching subsequences is not given by the random walk formula $1/(q-p)$, but elementary probability theory shows that $A = 1/(1-p)$. Using these various results, we get $K^* = [(1-p)]^2/[\{(1/p)-1\}] = pq$, and inserting this value in (25) we immediately recover (20).

The general BLAST theory covers cases more complicated than those described, by considering, first, DNA sequences of unequal length and all possible alignments between these sequences, and second, Markov dependence between successive nucleotides in the DNA. As noted above, the full theory is a *tour de force* of applied probability theory. Apart from this, given that BLAST is used perhaps hundreds of times each day, it is undoubtedly the most widely used, and significant, statistical method used in DNA sequence analysis.

**4. Finding disease genes – linkage analysis.** Linkage analysis, used now principally to find disease genes, has gone through several phases during the last eighty years.

First, the location of a major gene for hemophilia (and for other sex-linked genes) was found very early on. For example, it was noted even in historical times that hemophilia occurs much more often in men than in women, and thus from the genetic point of view, the gene for hemophilia is almost certainly on the $X$ chromosome.

This straightforward sort of observation led to the next approach to linkage analysis, namely case/control methods. These also operate through the concept of association. This approach makes fundamental use of "marker" genes, that is, genes at "marker" loci, having two essential properties: (i) the location of any marker locus on some chromosome is known, and (ii) one can tell the allelic types of the two marker genes any given individual has at a given marker locus.

In case/control methods we compare the marker locus genetic make-up of cases - individuals having the disease in question, and controls - individuals free of the disease. If the marker locus genotypes of these two groups differ significantly, it might be thought that the disease gene is closely linked to this marker. However this inference is not justified if population stratification can occur, since genetic associations can arise through such stratification. To take an extreme example, the population sampled might consist of a mixture of Swedes and Italians. Some given marker gene might occur at higher frequency among Swedes than among Italians, and the disease might also occur at higher frequency among Swedes than Italians. However, this does not necessarily imply linkage of the marker to the disease. When this argument became well-known, case-control methods fell into some disrepute, and interest moved to the third type of linkage analysis, the affected sib pair, or more generally affected relative pair, method. For the sib case, this method uses sharing properties of the numbers of transmissions of a marker gene to two affected sibs, in a manner described below.

Affected sib-pair methods also use some marker locus "M", of known location, and some putative disease locus "D", of unknown location. In formal statistical terms, the aim is to test the null hypothesis that the marker locus "M" is unlinked to the disease locus "D", against the alternative hypothesis that the marker locus "M" is linked to the disease locus "D" (the interesting case being "closely linked").

The method operates by considering sharing properties of marker genes among affected sib pairs. If "M" and "D" loci are unlinked, there will be no excess over random expectation of sharing of marker genes by affected sibs, apart from those caused by random statistical fluctuations. If "M" and "D" loci are linked, there will be on average an excess over random expectation of sharing of marker genes by affected sibs.

This argument if best illustrated by considering the case of a rare, recessive disease. Both sibs are affected, so both received the disease gene $D$ from both of their (most likely heterozygous $Dd$) parents. Thus since they share the disease gene passed on from both parents, they tend also to share the same marker genes at marker loci that closely linked to the disease locus. Testing for linkage using sharing marker locus data derives from this observation, and the testing procedure is a simple binomial test.

This test is typically done simultaneously with many different marker loci, scattered over the entire genome. However, testing many marker loci simultaneously leads to a high experimentwise Type I error. This difficulty is substantially resolved by noting that there is a high correlation in sharing characteristics between neighboring marker loci. It turns out that the degree of sharing, as one goes along any chromosome not having the disease locus on it, is described exactly by an Ornstein-Uhlenbeck process (Feingold et al., 1993). The test statistic used is the degree of sharing at the "most sharing" marker locus, and the distribution of this test statistic is found from the distribution of the maximum of an Ornstein-Uhlenbeck process over a given interval. We do not enter into the details of this procedure here.

There are two problems with affected sib-pair methods. First, the method requires data from families with at least two affected sibs, and such families are comparatively rare. Second, there is "too much" sharing in sibs, affected or otherwise. Sibs on average share about half their genetic material, and this creates significant "noise" in a procedure which tests for a significant excess in sharing. This problem is particularly acute for complex diseases, where the sharing behavior between affected sibs can be quite complex.

For these reasons, my colleagues R.S. Spielman, R.E. McGinnis and I (Spielman et al. 1993) introduced a new form of testing procedure, not relying on sharing methods, named the transmission/disequilibrium test (TDT). This is a new, fourth, form of linkage analysis. The test, and its statistical properties, are described in the remainder of this section.

The aim of the TDT is to combine the best features of association and family-based tests. The TDT can be thought of as a family-based case/control association procedure, except that the "control" with whom any affected individual is compared is not some random unaffected person, but is, rather, a "non-person" control created by the marker locus genes "thrown away" by the parents when the affected sib was conceived. For example, if the father of an affected child has the genotype $M_1M_2$ at a marker locus and the mother has genotype $M_3M_4$, and if the child himself has genotype $M_1M_4$, then this "non-person" control has genotype $M_2M_3$ at this locus. This explains the fact that the test is "family-based."

The TDT has three advantages over case/control and affected sib pair methods. First, unlike the case/control method, it overcomes the problem of population stratification by bringing the control "within the family". Second, unlike affected sib pair methods, the test requires only one affected child in the family. Finally, unlike affected sib-pair methods, it works well (for reasons discussed later) for complex diseases.

The TDT procedure of comparing cases and the artificial "controls" boils down to a comparison of what gene is transmitted, what gene is not transmitted, by the parent of an affected child. If there are two alleles, $M_1$ and $M_2$, at the marker locus, this implies that the data analyzed by the TDT procedure will be as in Table 1.

When there is no population stratification, then even if the mode of inheritance of the disease is known, there are four unknown parameters defining the probabilities of the four cells in this data table. Thus, for example, for a recessive disease, the probabilities $P_{11}, P_{12}, P_{21}$ and $P_{22}$ for these four cells are as given in (26), where $q$ and $p$ and respectively are the frequencies of $M_1$ and the disease gene, $\delta$ is the so-called "coefficient of association" between marker and disease loci and $\theta$ is the recombination fraction - a measure of the distance - between marker and disease loci.

$$\begin{aligned}
P_{11} &= q^2 + \delta q/p, \\
P_{12} &= q(1-q) + \delta(1-\delta-q)/p, \\
P_{21} &= q(1-q) + \delta(\theta-q)/p, \\
P_{22} &= (1-q)^2 - \delta(1-q)/p.
\end{aligned} \tag{26}$$

When population stratification exists there are many unknown parameters. Thus, for example, for a recessive disease, the probabilities $P_{11}, P_{12}, P_{21}$ and $P_{22}$ for these four cells are as given in (27), where the summation is over different strata in the population, $\alpha_i$ is the proportion of the population in

TABLE 1

Combinations of transmitted and nontransmitted marker alleles $M_1$ and $M_2$ among $2n$ parents of $n$ affected children.

| Transmitted | Non-transmitted allele | | |
|---|---|---|---|
| Allele | $M_1$ | $M_2$ | Total |
| $M_1$ | $n_{11}$ | $n_{12}$ | $n_{11}+n_{12}$ |
| $M_2$ | $n_{21}$ | $n_{22}$ | $n_{21}+n_{22}$ |
| Total | $n_{11}+n_{21}$ | $n_{12}+n_{22}$ | $2n$ |

stratum $i, q_i$ and $p_i$ and are the frequencies of $M_1$ and the disease gene in stratum $i$, and $\delta_i$ is the coefficient of association in stratum $i$.

$$\begin{aligned}
P_{11} &= \sum_i \alpha_i p_i[p_i\{q_i^2+\delta_i q_i\}]/\sum_i \alpha_i p_i^2,\\
P_{12} &= \sum_i \alpha_i p_i[p_i\{q_i(1-q_i)+\delta_i(1-\theta-q_i)\}]/\sum_i \alpha_i p_i^2,\\
P_{21} &= \sum_i \alpha_i p_i[p_i\{q_i(1-q_i)+\delta_i(\theta-q_i)\}]/\sum_i \alpha_i p_i^2,\\
P_{22} &= \sum_i \alpha_i p_i[p_i\{(1-q_i)^2-\delta_i(1-q_i)\}]/\sum_i \alpha_i p_i^2.
\end{aligned} \tag{27}$$

Despite the complexity of these probabilities and the impossibility of estimating the various parameters involved in the expressions in (27), it is straightforward to use the data in Table 1 to test for linkage between disease and marker loci. This is so because, when disease and marker loci are unlinked (i.e. when the null hypothesis $\theta = 1/2$ is true), no matter what the values of the remaining parameters may be, it is always true that $P_{12} = P_{21}$. Thus the test of the null hypothesis that disease and marker loci are unlinked reduces to a test of the hypothesis $P_{12} = P_{21}$. Standard Neyman-Pearson statistical theory shows that the appropriate test statistic is

$$(n_{12} - n_{21})^2/(n_{12} + n_{21}), \tag{28}$$

and that under the null hypothesis this has approximately a chi-square distribution with one degree of freedom. Expression (28) is the TDT statistic. Note that it uses only the data values $n_{12}$ and $n_{21}$. While this fact derives from Neyman-Pearson theory, it has an obvious genetical interpretation: these data values correspond to heterozygous ($M_1M_2$) parents, and only heterozygous parents can give information about linkage. Thus data from homozygous parents should not even be collected, since they will not be used in the test.

We make two observations concerning the use of the TDT statistic (28) as a test statistic of linkage between a marker and a disease. First, the main reason for introducing the statistic was to provide an association-based test which is not affected by population stratification. Now population stratification introduces association into the population, and this is measured by the quantities $\delta_i$ in (27). But it is easy to see from (27) that the larger the association in the population, the more powerful the test of the null hypothesis $\theta = 1/2$ becomes. (This can be seen from (27) by noting that if $\delta_i = 0$, then $P_{12} = P_{21}$, no matter what the value of $\theta$, so that in this case the TDT test has no power.) Thus not only is the TDT test not affected by population stratification and the resulting association, it can in some cases take advantage of the presence of association in increasing the power of the test (Ewens and Spielman, 1995).

The second observation concerns the claim, made above, that for marker loci very close to the disease locus, the TDT often provides a very powerful test of linkage. The reason for this is best explained by considering one example. The disease allele will have arisen by mutation some time in the past, perhaps several thousand years ago, on a chromosome bearing some marker allele, say $M_1$, as a marker locus "M". The disease gene has been passed on to a collection of affected individuals in the present generation. If the marker locus $M$ is extremely tightly linked to the disease locus, there will still be an excess of the $M_1$ allele among affected individuals, leading to an association between this allele and the disease allele. However, for a marker locus not tightly linked to the disease locus, this association will essentially have broken down, through the process of recombination, between the time of the initial mutation and the present generation. Thus the TDT, which relies for its operation on the presence of association, will pick up markers close to the disease locus (which are of course the markers of interest), but not others. For sufficiently close markers, there will be a strong association between marker and disease, and the TDT will tend to be significant. Of course, because there might be multiple initial mutational origins of a disease gene, at different disease loci, this argument is weakened. Nevertheless, the TDT might well pick up several of these disease loci.

TABLE 2

Combinations of transmitted and nontransmitted marker alleles $M_1, M_2, ..., M_k$ among $2n$ parents of $n$ affected children.

| | | Non-transmitted allele | | | | |
|---|---|---|---|---|---|---|
| | | $M_1$ | $M_2$ | .......... | $M_k$ | Total |
| | $M_1$ | $n_{11}$ | $n_{12}$ | | $n_{1k}$ | $n_{1.}$ |
| Transmitted allele | $M_2$ | $n_{21}$ | $n_{22}$ | | $n_{2k}$ | $n_2$ |
| | | | | .......................... | | |
| | $M_k$ | $n_{k1}$ | $n_{k2}$ | | $n_{kk}$ | $n_{k.}$ |
| | Total | $n_{.1}$ | $n_{.2}$ | | $n_{.k}$ | $2n$ |

All the above relates to the case where there are only two marker alleles, ($M_1$ and $M_2$). Since in many cases there are more than two marker alleles, it is necessary to generalize to the "$k$ marker alleles" case. Here the data will be in the form of Table 2. Note that in any test using the data in this table, homozygous parents should be simply ignored - as with the statistic (28), they will not be used in a test of linkage.

Various possible test statistics extending the TDT statistic (28) have been proposed using the data in this table. The (null) hypothesis that disease and marker loci are unlinked implies that the mean value of $n_{ij}$ is the same as that of $n_{ji}$, for all $(i, j)$ combinations. Thus one possible test for linkage between disease and marker loci is to test for symmetry in the data matrix of Table 2. However, for the case of $k$ marker alleles, this test has $k(k-1)/2$ degrees of freedom (df). Thus, use of such a test runs the risk of a "swamping" effect: one or a few markers with a strong effect might not be detectable in a global test that includes many markers with no effect or only a small effect. It is therefore agreed, on the whole, that such a global test should not be used.

At the next level of simplification, consider tests that focus on the data provided by the row totals $n_{1.}, ..., n_{k.}$ and the column totals $n_{.1}, ..., n_{.k}$ in

Table 2. These tests compare the number of transmissions $n_{i.}$ of the allele $M_i$ from heterozygous parents with the number of transmissions of an allele other than $M_i$ from heterozygous parents whose genotype contains $M_i$. The basis of these tests is that when the hypothesis of no linkage between disease and marker loci is true, the mean value of $n_{i.}$ is the same as that of $n_{.i}$, for all $i (i = 1, 2, .., k)$, and the test is in effect a test of the equality of these two means. There are two $k$-allele test statistics available which reduce exactly to the TDT statistic (28). The first of these is the generalized TDT (GTDT) statistic of Schaid (1996). Using this statistic one first calculates, for $i = 1, 2, ..., k$, the quantities $d_i$, defined by $d_i = n_{i.} - n_{.i}$. The sum of these quantities is necessarily zero, so that without loss of information one ignores one arbitrarily chosen marker allele (say allele $k$) and forms the vector $\mathbf{d}' = (d_1, d_2, ..., d_{k-1})$.

If the null hypothesis that disease and marker loci are unlinked is true, the estimate of the variance of $d_i$ is $n_{i.} + n_{.i} - 2n_{ii}$ and the estimate of the covariance between $d_i$ and $d_j$ is $-(n_{ij} + n_{ji})$. These variance and covariance estimates are formed into a matrix $V$ and the GTDT test statistic, defined as GTDT $= \mathbf{d}'\mathbf{V}^{-1}\mathbf{d}$, is then calculated. Note that, as required, this statistic does not use the values $n_{11}, n_{22}, ..., n_{kk}$, since these values cancel out in the definition of $\mathbf{d}'$ and $\mathbf{V}$. Under the null hypothesis of no linkage between disease and marker loci, the GTDT statistic has asymptotically a chi-square distribution with $k-1$ df, and thus can be used as a test of linkage by referring the observed value of the statistic to tables of significance points of this distribution. This statistic is well-known Stuart statistic (Agresti, 1990).

Although the GTDT statistic is possibly the most natural generalization of the TDT statistic, its calculation requires the inversion of a large and possibly sparse matrix, and indeed the inverse matrix might not exist for small data sets. It is thus useful to have a test statistic that is very similar to GTDT but which does not involve inversion of a matrix. Such a statistic was proposed by Spielman and Ewens (1996). This statistic is $W = \{(k-1)/k\} \sum_i (n_{i.} - n_{.i})^2/(n_{i.} + n_{.i} - 2n_{ii})$. As with the GTDT statistic, this statistic also does not use the data values $n_{11}, n_{22}, ..., n_{kk}$. If the values of the quantities $n_{ij} + n_{ji}$ are not too dissimilar, it also has a distribution close to chi-square with $k = 1$ df under the null hypothesis of no linkage. Like the GTDT statistic, it also reduces to the two-allele TDT statistic when $k = 2$.

The two tests described above using all marker alleles symmetrically have $k - 1$ df and thus run the risk of swamping a real association with one specific marker allele. A statistic designed to avoid this problem is the maxTDT statistic of Schaid (1996). For each $i, (i = 1, 2, ..., k)$ we lump all alleles other than allele $i$ as "non-i" and compute a "two-allele" TDT statistic as prescribed in (28). We then choose the largest of the $k$TDT statistics so formed as the test statistic, denoting this statistic max TDT. In terms of the entries in Table 2, this statistic is the largest, as

$i$ takes successively the values $1, 2, ...k$, of the "two-allele" TDT statistics $(n_{i.} - n_{.i})^2/(n_{i.} + n_{.i} - 2n_{ii})$. This test statistic also reduces to the TDT statistic when $k = 2$.

The maxTDT statistic has 1 df, and thus largely avoids the swamping effect mentioned above. However one may not use chi-square tables to test for its significance, since such a deliberately chosen largest TDT statistic does not have a chi-square distribution. Nor are simple Bonferroni corrections for the significance points completely accurate for this test, because there is a simple linear constraint between the terms that are squared in the numerator of each such statistic. The significance points of the maxTDT statistic have been found by simulation and significance values are available.

Several points of statistical interest arise from these considerations. What is the best test statistic, of those we have considered, for testing for linkage? Would a permutation test be better, particularly if the asymptotic distribution of the test statistic used is not clear? As an associated question, when should we ignore formal Neyman-Pearson theory (in favor of principles derived from purely genetical arguments) as not being helpful? On a more concrete point, what is the distribution of maxTDT? These questions, and similar questions in physical methods using clones and anchors, and of analyzing DNA sequence data, are as yet unanswered.

## REFERENCES


[1] AGRESTI, A., *Categorical Data Analysis*, Wiley, New York, 1990.

[2] ARRATIA, R.E. LANDER, S. TAVARE AND M.S. WATERMAN, Genomic mapping by anchoring random clones: a mathematical analysis, *Genomics*, 11:806–827, 1991.

[3] BOTSTEIN, D., R.L. WHITE, M. SKOLNIK AND R.W. DAVIS, Construction of a genetic linkage map in man using restriction fragment length polymorphisms, *Am. J. Hum. Genet*, 32:314–331, 1980.

[4] EWENS, W.J., C.J. BELL, P.J. DONNELLY, R. DUNN, E. MATALLANA AND J.R. ECKER, Genome mapping with anchored clones: theoretical aspects, *Genomics*, 11:799–805, 1991.

[5] EWENS, W.J. AND R.S. SPIELMAN, The transmission/disequilibrium test: history, subdivision and admixture, *Am. J. Hum. Genet*, 57:455–464, 1995.

[6] FEINGOLD, E., P.O. BROWN AND D. SIGMUND, Gaussian models for genetic linkage analysis using complete high-resolution maps of identity by descent, *Amer. J. Hum. Genet.*, 53:234–251, 1993.

[7] FELLER, W., *An Introduction to Probability Theory and its Application*, Wiley, New York, 1957.

[8] KARLIN, S. AND S.F. ALTSCHUL, Methods for assessing the statistical significance of molecular sequence features by using general scoring schemes, *Proc. Nat. Acad. Sci.*, 87:2264–2268, 1990.

[9] KARLIN, S. AND S.F. ALTSCHUL, Applications and statistics for multiple high-scoring segments in molecular sequences, *Proc. Nat. Acad. Sci.*, 90:5873–5877, 1993.

[10] KARLIN, S. AND A. DEMBO, Limit distributions of maximal segmental score among Markov-dependent partial sums, *Adv. Appl. Prob.*, 24:113–140, 1992.

[11] OLSON, M.V., L. HOOD, C. CANTOR AND D. BOTSTEIN, A common language for physical mapping of the human genome, *science*, 245:1434–1435, 1989.

[12] SCHAID, D.J., General score tests for associations of genetic markers with disease using cases and their parents, *Genet. Epidemiol.*, 13:423–450, 1996.

[13] SCHBATH, S., Coverage processes in physical mapping by anchoring random clones, *J. Comp Biol*, 4:61–82, 1997.

[14] SCHLESSINGER, D., Yeast artificial chromosomes: tools for mapping and analysis of complex genomes, *Trends in Genetics*, 6:248–259, 1990.

[15] SPIELMAN, R.S., R.E. MCGINNIS AND W.J. EWENS, Transmission test for linkage disequilibrium: the insulin gene region and insulin-dependent diabetes mellitus, *Am. J. Hum. Genet.*, 52:506–516, 1993.

[16] SPEILMAN, R.S. AND W.J. EWENS, The TDT and other family-based tests for linkage disequilibrium and association, *Am. J. Hum. Genet*, 59:983–989, 1996.

[17] WILLIAMS, J.G., A.R. KUBELIK, K.J. LIVAK, J.A. RAFALSKI AND S.V. TINGEY, DNA polymorphisms amplified by arbitrary primers are useful as genetic markers, *Nucleic Acids Res.*, 18:6531–6535, 1990.

# THE AFFECTED-PEDIGREE-MEMBER METHOD REVISITED UNDER POPULATION STRATIFICATION

CHI GU*, MICHAEL B. MILLER†, THEODORE REICH‡, AND D.C. RAO§

**Abstract.** The effect of population stratification on the Affected-Pedigree-Member (APM) method is explored herein. Formulae for bias of means and variances of APM distributions are derived assuming no mating across subpopulations. We find that the choice of a weighting function of marker frequency greatly affects the test result in the presence of population stratification. In particular, when the weight function $f(p) = 1/p$ is applied, the mean of APM test score is not biased and and the bias of variance can be estimated as a dependent variable of the distribution of actual marker allele frequencies in subpopulations. A method to correct the APM test score for stratification is proposed to produce conservative significance levels. Computer simulations are carried out to assess biases of APM due to population stratification when various weight functions are used.

**Key words.** APM method, IBS, IBD, linkage analysis, population stratification.

## 1. Introduction.

The Affected-Pedigree-Member (APM) method is an inexpensive tool for linkage analysis whenever IBS (identical by state) data are available (marker data on parents are ignored). However, as several authors discovered ([1], [4]), the APM method tests for a composite hypothesis: no linkage and correctly specified marker allele frequencies. Its well known sensitivity to the misspecification of the marker frequencies may lead to spurious detection of linkage.

In this paper we study a special case of misspecification — when population stratification occurs. In this situation, allele frequencies are misspecified in subpopulations even though they maybe correctly estimated for the whole population. We show that, under the null hypothesis of no linkage, the APM test score is biased toward linkage. The deviation depends on the stratification fractions, the marker allele frequencies in the subpopulations, the pedigree structures, and most importantly, on the weight functions being used. We show analytically that when the weight function $f(p) = 1/p$ is used, the mean is unbiased and we can estimate the bias of variance from the distribution of marker frequencies in subpopulations. A way to correct the APM test score is then proposed to produce conservative significance levels accounting for stratification effect. We then proceed to investigate the dependence of biases on different pedigree structures and show that for weight function $f(p) = 1/p$, more distant affected relative pairs on the

*Division of Biostatistics, Washington University, School of Medicine, St. Louis, MO 63110-1093.

†Department of Psychology 210 McAlester Hall University of Missouri–Columbia Columbia, MO 65211.

‡Dept. of Psychiatry and Genetics, Washington University, School of Medicine, St. Louis, MO 63110-1093.

§Division of Biostatistics, Psychiatry and Genetics, Washington University, School of Medicine, St. Louis, MO 63110-1093.

pedigrees results in APM test scores that are less sensitive to the underlying stratification.

The effect of population stratification on the APM test scores is presented in §2, so does the formulae for deviations of expectations and variances. At the end of §2 we give results of simulation studies to show the stratification effect on the significance level of APM test. In all calculations, affected parent-offspring pairs are excluded as suggested in the literature ([4, 2, 3]).

**2. APM score in stratified populations.** To study the effect of population stratification on APM statistics, we need to compare the APM scores calculated with the true marker allele frequencies in the subpopulation and with the frequencies estimated for the stratified population as a whole.

Assume that the overall population is a mixture of $P$ different subpopulations $P_1$, $P_2$, ..., $P_P$ with proportions $m_1$, $m_2$, ..., $m_P$, $(0 < m_i < 1)$, and no mating cross subpopulations. We also assume that all subpopulations are homogeneous at the disease susceptibility locus and are at Hardy-Weinberg equilibrium with the marker locus within each subpopulation. Subpopulations having the same distributions of marker allele frequencies are treated as a single subpopulation. Denote the marker gene frequencies by $p_{k,i}$ for allele $k$ in population $P_i$. Let $\bar{p}$ denote the pooled gene frequency for the whole population, that is

$$\bar{p}_k = \sum_i m_i p_{k,i} \ ,$$

$i = 1, ..., P$, $k = 1, ..., N$, will always be different from the true values in at least one of the subpopulations. We will use subscript "$Fm$" to denote quantities calculated for a single pedigree in the sample, and subscript "$IJ$" to denote quantities for an affected relative pair "I" and "J". Using the same notation as in Weeks and Lange (1988), with slight modification, $Z_{IJ}$ denotes weighted conditional expectation of IBD sharing between relative pair $I$ and $J$; $Z_{Fm}$ denotes the score for overall IBD sharing on pedigree $Fm$. The APM test statistic $T$ is defined as the normalized sum of $Z_{Fm}$ for all families in the sample (See formulas (3), (4), (7), (8) in [5]). We will refer to $Z_{Fm}$ by APM score or simply "Z", and to $T$ by APM test statistic.

It is mathematically straightforward to estimate the actual mean and variance of the APM score under the null hypothesis of no linkage for the stratified population. One only needs to observe the fact that probability of an event now is decomposed as a sum of conditional probabilities (conditioned on whether the family is in subpopulation $P_i$). Therefore,

$$\begin{aligned} E(Z_{\mathrm{Fm}}) &= \sum_{\forall Z_0} Z_0 \Pr(Z_{\mathrm{Fm}} = Z_0) \\ &= \sum_{\forall Z_0} Z_0 \sum_i \Pr(Z_{\mathrm{Fm}} = Z_0 |\ \mathrm{Fm} \in P_i) \Pr(\ \mathrm{Fm} \in P_i) \end{aligned}$$

$$= \sum_i m_i E_i(Z_{\mathrm{Fm}}), \tag{2.1}$$

where $E_i(Z_{\mathrm{Fm}})$ is the mean of the APM score for subpopulation $P_i$, and may be computed via formulas (see [5]):

$$E_i(Z_{\mathrm{Fm}}) = \sum_{I<J} E_i(Z_{IJ}), \tag{2.2}$$

where

$$E_i(Z_{IJ}) = \Phi_{IJ} \sum_{k=1}^{n} p_{k,i} f(p_{k,i}) + (1 - \Phi_{IJ}) \sum_{k=1}^{n} p_{k,i}^2 f(p_{k,i}) \tag{2.3}$$

By the same token, the variance of the APM score for the overall stratified population equals:

$$\begin{aligned} V(Z_{\mathrm{Fm}}) &= E(Z_{\mathrm{Fm}}^2) - E(Z_{\mathrm{Fm}})^2 \\ &= \sum_{\forall Z_0} Z_0^2 \Pr(Z_{\mathrm{Fm}} = Z_0) - E(Z)^2 \\ &= \sum_i m_i E_i(Z_{\mathrm{Fm}}^2) - E(Z)^2 \end{aligned} \tag{2.4}$$

where $E_i(Z_{\mathrm{Fm}}^2)$ may be replaced by $(V_i(Z_{\mathrm{Fm}}) + E_i(Z_{\mathrm{Fm}})^2)$ if variances and expectations are available for the subpopulations.

**2.1. Deviations of means and variance of Z.** Formulas 2.1–4 use the actual marker allele frequencies in the subpopulations which are not known in practice. Nonetheless, they may be used to estimate deviations of means and variances between the true distribution of APM scores and that derived from the misspecified marker frequencies pooled over the whole population. From the above formulas, we may calculate the deviations of means and variances between the true APM score for the stratified population and that derived from the misspecified marker allele frequencies for the whole population. By carefully studying these deviations, we will be able to estimate the correct significance level of a APM test statistic calculated with the pooled marker frequencies.

Let us denote the mean and variance of the APM score calculated with the pooled marker allele frequencies for the whole population by $\bar{E}(Z)$ and $\bar{V}(Z)$. Then we have, for each affected relative pair on a pedigree,

$$\bar{E}(Z_{IJ}) = \Phi_{IJ} \sum_k p_k f(p_k) + (1 - \Phi_{IJ}) \sum_k {p_k}^2 f(p_k) \tag{2.5}$$

The difference between (2.3) and (2.5) is the marker frequencies used.

Now we may write the deviation between the "real" mean $E(Z)$ and the "estimated" mean $\bar{E}(Z)$ as following:

$$\Delta E(Z_{\mathrm{Fm}}) = \sum_{I<J} \Delta E(Z_{IJ}) \tag{2.6}$$

and,

$$\begin{aligned}\Delta E(Z_{IJ}) &= E(Z_{IJ}) - \bar{E}(Z_{IJ}) = \sum_i m_i E_i(Z_{IJ}) - \bar{E}(Z_{IJ}) \\ &= \sum_i m_i[\Phi_{IJ} \sum_k p_{k,i} f(p_{k,i}) + (1-\Phi_{IJ}) \sum_k p_{k,i}{}^2 f(p_{k,i})] \\ &\quad -[\Phi_{IJ} \sum_k p_k f(p_k) + (1-\Phi_{IJ}) \sum_k p_k{}^2 f(p_k)] \\ &= \sum_k \left\{ \Phi_{IJ} \left( \sum_i m_i [p_{k,i} f(p_{k,i})] - [\sum_i m_i p_{k,i}] f(\sum_i m_i p_{k,i}) \right) + \right. \\ &\quad \left. (1-\Phi_{IJ}) \left( \sum_i m_i [p_{k,i}{}^2 f(p_{k,i})] - [\sum_i m_i p_{k,i}]^2 f(\sum_i m_i p_{k,i}) \right) \right\} \\ &\equiv \sum_k \left( \sum_i m_i F_{IJ}(p_{k,i}) - F_{IJ}\left( \sum_i m_i p_{k,i} \right) \right)\end{aligned} \tag{2.7}$$

where function $F_{IJ}(x) = \Phi_{IJ} x f(x) + (1-\Phi_{IJ}) x^2 f(x)$.

By virtue of formula (2.7), we see that the sign and value of $\Delta E(Z_{IJ})$ now is determined by the concavity of $F_{IJ}(x)$. Simple algebraic calculation (Table 1) shows that, with $0 < \Phi_{IJ} < 1$, for weight function $f(p) = 1$ it is concave up, for $f(p) = 1/p$ it is linear, and for $f(p) = 1/\sqrt{p}$ it is indefinite. Therefore, when using weight function $f(p) = 1/p$, the population stratification puts no bias on the APM score, i.e. the mean of the score is the same as the correct one. If the weight function $f(p) = 1$ is used, the difference term in each of the summand in (2.7) is positive, thus the deviation is always positive. i.e. the expectation of the estimated APM score is always smaller than the true one. If the weight function $f(p) = 1/\sqrt{p}$, the sign of deviation is undetermined.

The deviation of the variance is not as simple to estimate analytically. From formula (2.4), we may write the deviation as:

$$\begin{aligned}\Delta V(Z) &= V(Z) - \bar{V}(Z) \\ &= [E(Z^2) - \bar{E}(Z^2)] + [\bar{E}(Z)^2 - E(Z)^2] \\ &= \sum_i m_i \left[E_i(Z^2) - \bar{E}_i(Z^2)\right] - \Delta E(Z)(E(Z) + \bar{E}(Z)) \\ &= \sum_{\substack{I<J \\ K<L}} \left[ \sum_i m_i E_i(Z_{IJ} Z_{KL}) - \bar{E}(Z_{IJ} Z_{KL}) \right] - \Delta E(Z)(\bar{E}(Z) + E(Z))\end{aligned} \tag{2.8}$$

So the second part has the same sign as $-\Delta E(Z)$ ( $E(z)$ and $\bar{E}(z)$ always non-negative). To determine the sign of $\Delta V(Z)$ and estimate its absolute value, we need to study the difference of the second moments

$$\sum_i m_i E_i(Z_{IJ}Z_{KL}) - \bar{E}(Z_{IJ}Z_{KL}) \tag{2.9}$$

for all $I < J$ and $K < L$. The evaluation of $E_i(Z_{IJ}Z_{KL})$ etc. relies on the recursive calculation of the generalized kinship coefficients (see formula (6) in [5]). We list the 6 gene frequency factors and its corresponding generalized kinship coefficients in columns of Table 2. Each second moment is the sum of the products of these factors and its corresponding coefficients.

We denote by $\varphi_t$ the sum of one type of generalized kinship coefficients over all available affected relative pairs on a pedigree, where $t$ indicates row in Table 2. e.g.

$$\varphi_1 = \sum_{\substack{I<J \\ K<L}} \Phi[(G_I, G_J, G_K, G_L)] \tag{2.10}$$

Therefore, the second moment difference can be written as

$$\begin{aligned}
&\Delta E(Z^2) \\
&= \varphi_1 \left[ \sum_k \left( \sum_i m_i p_{ki} f(p_{ki})^2 - (\sum_i m_i p_{ki}) f(\sum_i m_i p_{ki})^2 \right) \right] + \\
&+ \varphi_2 \left[ \sum_i m_i \left( \sum_k p_{ki} f(p_{ki}) \right)^2 - \left( \sum_k (\sum_i m_i p_{ki}) f(\sum_i m_i p_{ki}) \right)^2 \right] + \\
&+ \varphi_3 \left[ \sum_k \left( \sum_i m_i p_{ki}^2 f(p_{ki})^2 - (\sum_i m_i p_{ki})^2 f(\sum_i m_i p_{ki})^2 \right) \right] + \\
&\varphi_4 \left[ \sum_{k,l} \left( \sum_i m_i p_{ki}^2 f(p_{ki}) p_{li}^2 f(p_{li}) - (\sum_i m_i p_{ki})^2 f(\sum_i m_i p_{ki}) \right. \right. \\
&\qquad \left. \left. (\sum_i m_i p_{li})^2 f(\sum_i m_i p_{li}) \right) \right] \\
&+ \varphi_5 \left[ \sum_k \left( \sum_i m_i p_{ki}^3 f(p_{ki})^2 - (\sum_i m_i p_{ki})^3 f(\sum_i m_i p_{ki})^2 \right) \right] + \\
&+ \varphi_6 \left[ \sum_i m_i \left( \sum_k p_{ki}^2 f(p_{ki}) \right)^2 - \left( \sum_k (\sum_i m_i p_{ki})^2 f(\sum_i m_i p_{ki}) \right) \right]
\end{aligned} \tag{2.11}$$

Notice that the sum of the generalized kinship coefficients only depends on the pedigree structure and the affection status on the pedigrees. Therefore, the sign of $\Delta E(Z^2)$ is related to the concavity of functions in

Table 3. In particular, for weight function $f(p) = 1/p$, using $\sum_i m_i = 1$, $\sum_k p_k = 1$ and $\sum_k p_{ki} = 1$, we see that summands in (11) are all equal to zero except the first one. For the first summand in (11), since function $\frac{1}{x}$ concave up on interval [0,1] we see that $\Delta E(Z^2) \geq 0$. Combining with the fact that $\Delta E(Z) = 0$ for $f(p) = 1/p$, we conclude that $\Delta V(Z) \geq 0$. The APM statistic calculated from the pooled marker frequencies is normalized with the incorrect variance terms. The bias of variance is

$$\sum_m w_m^2 (V(Z_{\mathrm{Fm}}) + \Delta V(Z_{\mathrm{Fm}})). \tag{2.12}$$

Thus the correct normalization should be

$$T = \frac{\sum_m w_k [Z_{\mathrm{Fm}} - E(Z_{\mathrm{Fm}})]}{\sqrt{\sum_m w_k^2 (V(Z_{\mathrm{Fm}}) + \Delta V(Z_{\mathrm{Fm}}))}} \to N(0,1). \tag{2.13}$$

**2.2. Adjust APM score for stratified populations.** We see in the previous subsection that the biases of both mean and variance of APM score depend on the weight function, the marker allele frequencies for each subpopulation, the degree of stratification and the pedigree structure. It will assist greatly to a genome wide search if one can estimate these deviation given small amount of information about these independent variables. We give such a procedure in this section. With some understanding of allele frequencies among different populations, on may use the weight function $f(p) = 1/p$, and estimate the maximum possible deviations in variance of the APM score. Thus a correction of APM test statistic calculated using the pooled marker gene frequencies becomes possible.

Let us take a loser look at the deviation of variances when the weight function is $f(p) = \frac{1}{p}$. Calculations in the previous subsection show that,

$$\Delta V(Z) = \varphi_1 \sum_k \left[ \sum_i m_i \frac{1}{p_{k,i}} - \frac{1}{\sum_i m_i p_{k,i}} \right]. \tag{2.14}$$

By virtue of the concavity of $f(x) = \frac{1}{x}$, we have that all summand in the above sum is non-negative, thus $\Delta V(Z) \geq 0$. Therefore the APM score we got for the stratified population is always inflated, resulting in a significance level that is higher than the actual one. See Figure-1 for an illustration of the relationship between the true underlying APM score and the the one using the estimated marker frequencies from the stratified population.

In practice the stratification fraction may not be available, neither the marker frequencies in the subpopulations. However, if the lower and upper boundaries of allele frequencies for the subpopulations are known, we may estimate the maximum possible deviation of variance by

$$\Delta_{\max} V(Z) = \varphi_1 \sum_k \left[ \frac{1}{\sqrt{p_{k\ \min}}} - \frac{1}{\sqrt{p_{k\ \max}}} \right]^2. \tag{2.15}$$

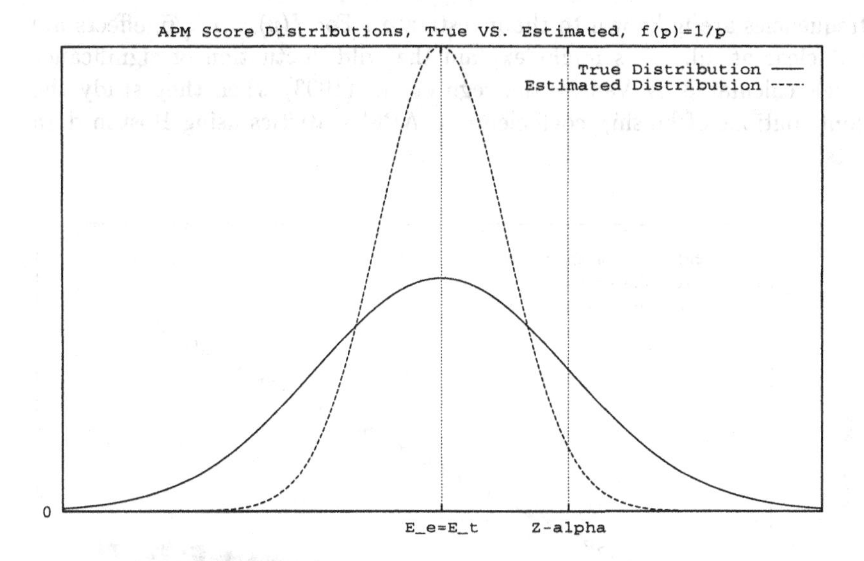


FIG. 1. *True VS. estimated scores using f(p)=1/p*

So we may adjust the APM test statistic calculated with the estimated allele frequencies by the following formula:

$$\bar{T} = \frac{\sum_m w_k[Z_{\text{Fm}} - E(Z_{\text{Fm}})]}{\sqrt{\sum_m w_k^2(V(Z_{\text{Fm}}) + \Delta_{\max} V(Z_{\text{Fm}}))}} \tag{2.16}$$

The significance level obtained for this adjusted test statistic is then always higher than the actual one thus eliminating the false positives. The difference in significance may be estimated by

$$\Delta\alpha = \Pr(N(0,1) > T) - \Pr(N(0,1) > \bar{T}) \tag{2.17}$$

We notice that if the correction is too conservative the power of the test could be reduced.

**2.3. Configuration of affection status on pedigrees.** From formulas (2.7), (2.11) and (2.8), we see how configuration of affection status on pedigrees would affect the bias of the APM score via the standard and generalized kinship coefficients. In particular, when weight function $f(p) = 1/p$ is used, since there is no bias for means the smaller $\varphi' s$ would result less bias of variance. Thus more distant affected relative pairs on the pedigrees make the APM score less sensitive to the underlying stratification.

For $f(p) = 1$, $\Delta E(Z)$ has a multiplicative factor $\sum_{I<J}(1 - \Phi_{IJ})$. So less distant relative pairs which produce smaller sum $\sum_{I<J}(1 - \Phi_{IJ})$ will generate less bias of APM means. However, the effect on the bias becomes less tractable if the degree of stratification and the distribution of marker

frequencies are unknown to the investigator. For $f(p) = 1/\sqrt{p}$, effects are not clear at all. This might explain the wild fluctuation of significance levels calculated by Van Eerderwegh et al. (1993) when they study the contributions of kinship coefficients to APM statistics using Boston data sets.

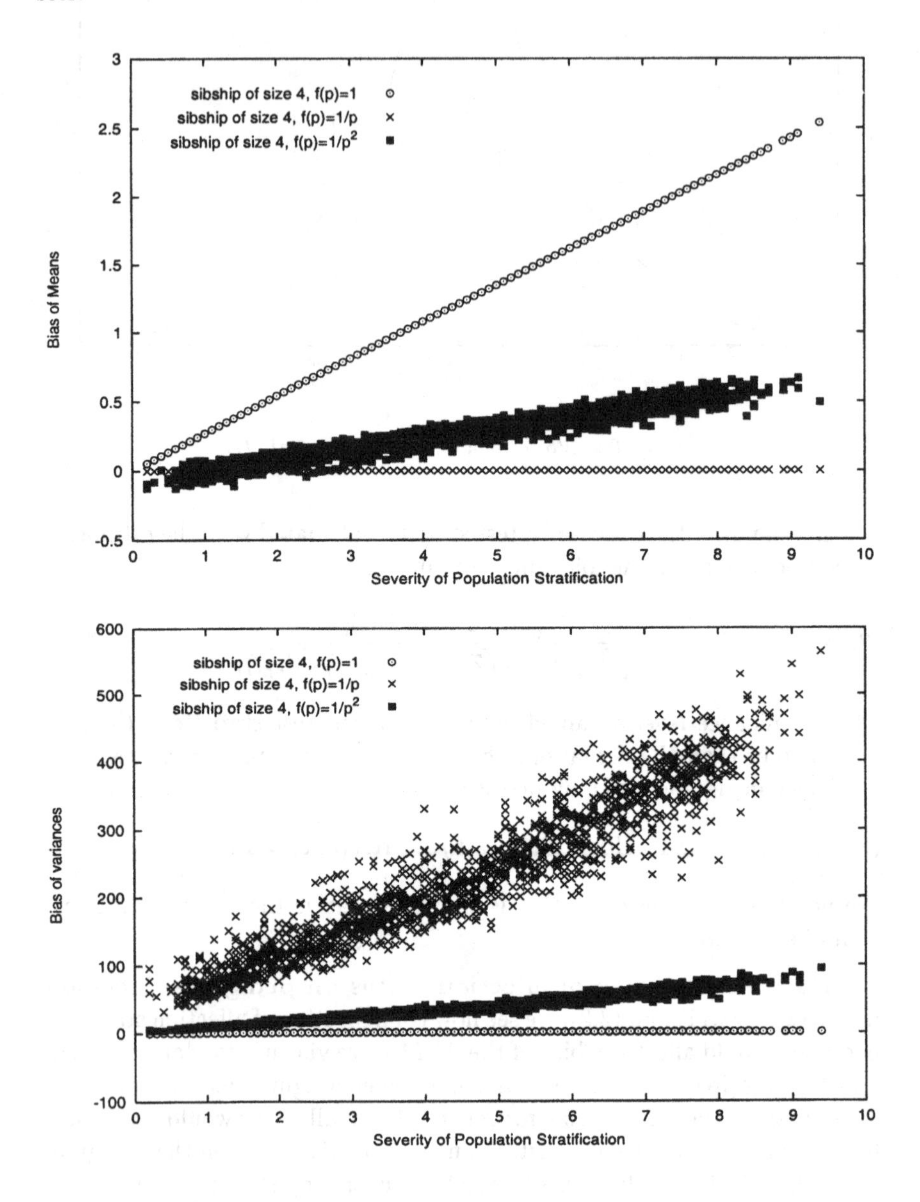


FIG. 2. *Bias of means (Top) and variances (Bottom) plotted against degree of stratification*

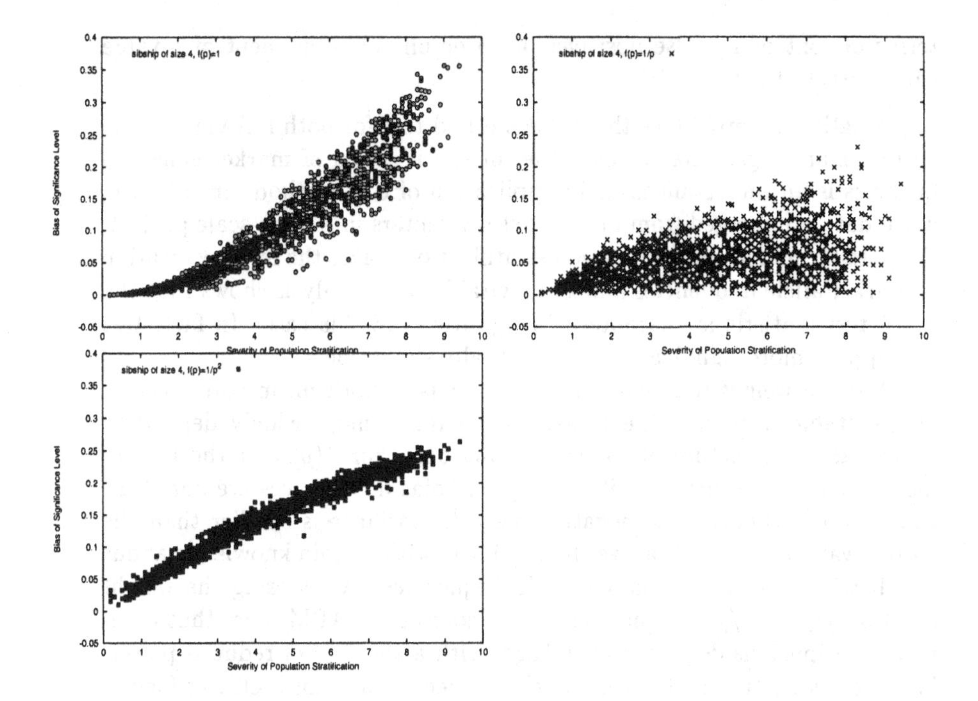


FIG. 3. *Bias of p-values against a nominal significance level* $\alpha = 0.001$, *using different weight functions*

**2.4. Simulation results.** To illustrate the effect of population stratification on APM test, we display here some results of computer simulation analysis.

We introduce a measure of degree of population stratification (in terms of marker allele frequencies), "$s$", defined by:

$$s = 100 \times (\text{weighted sum of squares error of the marker frequencies}).$$

Suppose each pedigrees in the sample has both parents and a sibship of 4 affected siblings. At a marker locus with 6 alleles which is completely linked to the disease trait, we then randomly assign values of marker frequencies in a pool of five subpopulations, and calculated bias of means and variances for the APM statistic. The results are shown for all three weight functions considered in Figure 1&2. The effect of these biases on the significance of APM test is shown in Figure 3, plotted separately for each weight function. It seems that the weight function $f(p) = 1/p$ has less sever effects on APM significance for a lot more cases than the other two weight functions, although the effects on the bias of p-values when population stratification is presented are more sensitive to the actual configurations of the subpopulations (see Figure 4). On the other hand, the bias in p-values is much less scattered for weight function $f(p) = 1/p^2$, across all the values of $s$ tested. This indicates that weight function $f(p) = 1/p^2$ may perform better in terms of controlling false positives, if an empirical adjustment of p-value can be worked out.

**3. Discussion.** Since the APM method test for both linkage and the correct marker gene frequencies, the misspecification of marker gene frequencies is an important issue in application of this method. In practice, when data are merged from different study centers in a large-scale project, the possible underlying population stratification makes the misspecification of marker allele frequencies almost inevitable. Our analysis shows that the APM test statistic is guaranteed being altered in this case. In fact, test will appear more significant than it actually should be.

For the weight function $f(p) = 1/p$, bias of both mean and variance are tractable, although the biases spread much more widely depending on the actual structure of the subpopulations. For $f(p) = 1$ the bias in means is always positive, while the signs of bias in variances are not clear. Chances are, they will be negative (i.e., the variance is smaller than the correct value). However, as we shown above, with certain knowledge about the distributions of the marker allele frequencies, by choosing the weight function $f(p) = 1/p$, one may be able to correct the APM score thus eliminate the spurious detection of linkage, with a sacrifice of reduced power. When the structure and marker allele frequencies are much clearer for the subpopulations, empirical adjustment of p-values will be possible and the weight function $f(p) = 1/p^2$ then should be used, which will guarantee the robustness of the adjustment across many subpopulation configurations. Configuration of affection status on pedigrees is also of importance when stratification is presented in the sample. If $f(p) = 1/p$ is used, more distant affected relative pairs would make a better design that is more robust against the stratification effect. In general, pedigrees with small $\varphi's$ will result in negative bias in variances and the bias in means will be estimable.

We did not consider the cases when some subpopulations have null alleles at the marker locus (i.e. $p_{ki} = 0$ for some $k$ and $i$), or when there is random mating among the subpopulations (panmixia). The extension of the results is straightforward but more laborious (e.g., with null allele, for $f(p) = 1/p$, $\Delta E(Z)$ is no longer zero but negative, and $\Delta V(Z)$ become less tractable). Often, investigators will see discrepancy between a non-significant IBD test score (which should not be sensitive to misspecification of allele frequencies) and a significant APM test statistic. To decide whether there is a real signal of genetic linkage or just noise due to population stratification, one needs to investigate the effect of population stratification on APM under the alternatives, i.e. when there is linkage. We will address these questions in a separate paper.

**Acknowledgments.** This work was supported in part, by USPHS grant MH14677 and NIGSM grant GM28719.

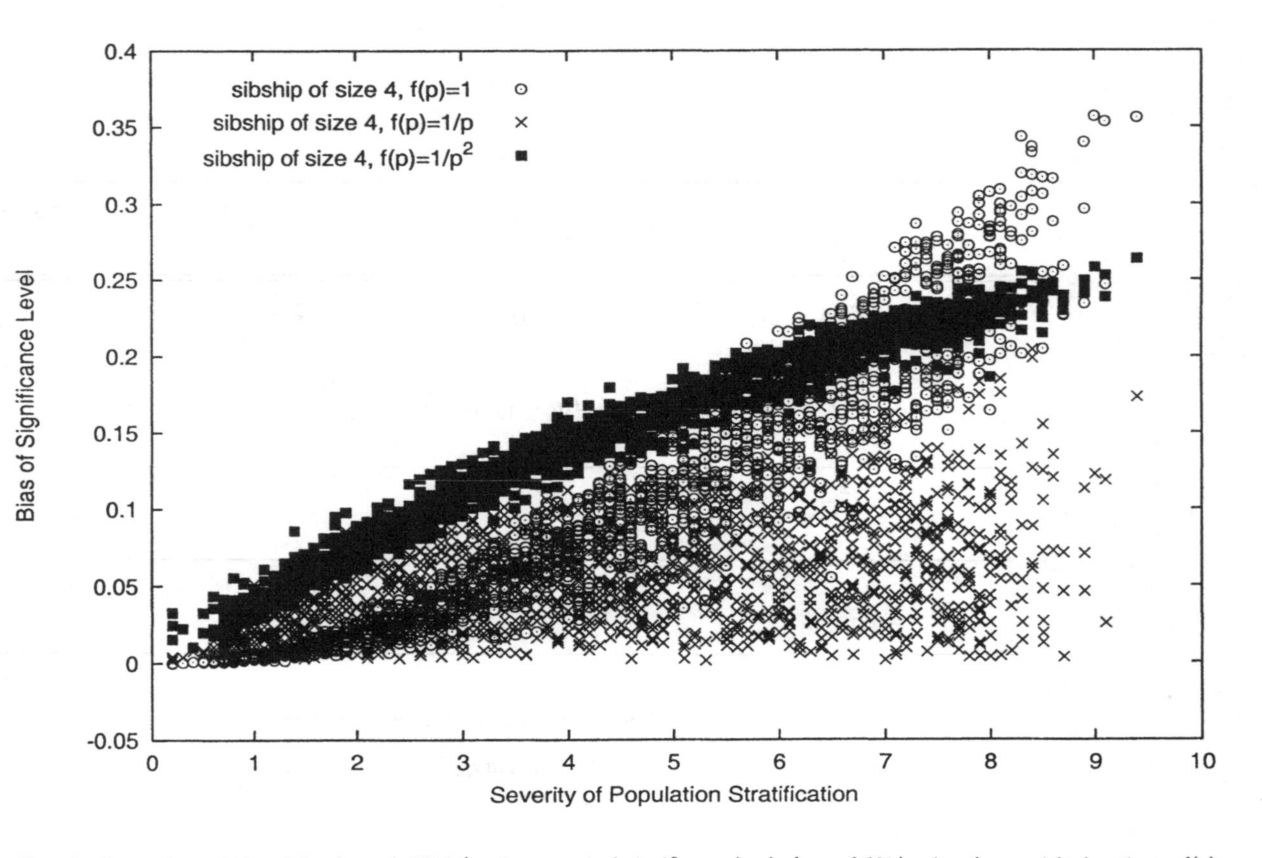


FIG. 4. *Comparison of bias of P-values of APM (w.r.t. a nominal significance level of* $\alpha = 0.001$*) using three weight functions:* $f(p) = 1, 1/p$*, and,* $1/p^2$*. Affected sibships of size 4 was used in the calculation.*

TABLE 1

*Deviation $\Delta E(Z_{IJ})$ and the concavity of $F(x)$*

| | $f(x) =$ | | |
|---|---|---|---|
| | $1$ | $\frac{1}{x}$ | $\frac{1}{\sqrt{x}}$ |
| $F(x)$ | $\Phi_{IJ}x + (1-\Phi_{IJ})x^2$ | $\Phi_{IJ} + (1-\Phi_{IJ})x$ | $\sqrt{x}(\Phi_{IJ} + (1-\Phi_{IJ})x)$ |
| $F''(x)$ | $2(1-\Phi_{IJ}) \geq 0$ | $0$ | $-\frac{1}{4\sqrt{x}}(3(1-\Phi_{IJ}) - \Phi_{IJ} - \frac{1}{x})$ |
| $\Delta E(Z_{IJ})$ | $\geq 0$ | $0$ | Indefinite |

TABLE 2

*Calculation of Second Moments*

| Marker Frequency Factor | Generalized Kinship Coefficient | |
|---|---|---|
| $[\sum_k p_k f(p_k)^2]$ | $\Phi[(G_I, G_J, G_K, G_L)]$ | $\varphi_1$ |
| $[\sum_k p_k f(p_k)]^2$ | $\Phi[(G_I, G_J)(G_K, G_L)]$ | $\varphi_2$ |
| $[\sum_k {p_k}^2 f(p_k)^2]$ | $\Phi[(G_I, G_J, G_K)(G_L)]$, $\Phi[(G_I, G_J, G_L)(G_K)]$, | $\varphi_3$ |
| | $\Phi[(G_I, G_K, G_L)(G_J)]$, $\Phi[(G_J, G_K, G_L)(G_I)]$, | |
| | $\Phi[(G_I, G_L)(G_J, G_K)]$, $\Phi[(G_I, G_K)(G_J, G_L)]$, | |
| $[\sum_k {p_k}^2 f(p_k)] \cdot [\sum_k p_k f(p_k)]$ | $\Phi[(G_I, G_J)(G_K)(G_L)]$, $\Phi[(G_I)(G_J)(G_K, G_L)]$ | $\varphi_4$ |
| $[\sum_k {p_k}^3 f(p_k)^2]$ | $\Phi[(G_I, G_K)(G_J)(G_L)]$, $\Phi[(G_I, G_L)(G_J)(G_K)]$, | $\varphi_5$ |
| | $\Phi[(G_J, G_L)(G_I)(G_K)]$, $\Phi[(G_J, G_K)(G_I)(G_L)]$, | |
| $[\sum_k {p_k}^2 f(p_k)]^2$ | $\Phi[(G_I)(G_J)(G_K)(G_L)]$ | $\varphi_6$ |

TABLE 3

*Signs of $\Delta V$ for Different Weight Functions*

| | | $f(x) =$ | | |
|---|---|---|---|---|
| | | $1$ | $\frac{1}{x}$ | $\frac{1}{\sqrt{x}}$ |
| $[\sum_k p_k f(p_k)^2]$ | $x f(x)^2$ | $1$ | $\frac{1}{x}$ | $1$ |
| $[\sum_k p_k f(p_k)]^2$ | $x f(x) y f(y)$ | $xy$ | $1$ | $\sqrt{x}\sqrt{y}$ |
| $[\sum_k {p_k}^2 f(p_k)^2]$ | $x^2 f(x)^2$ | $x^2$ | $1$ | $x^{\frac{3}{2}}$ |
| $[\sum_k {p_k}^2 f(p_k)] \cdot [\sum_k p_k f(p_k)]$ | $x^2 f(x) y f(y)$ | $x^2 y$ | $x$ | $x^{\frac{3}{2}}\sqrt{y}$ |
| $[\sum_k {p_k}^3 f(p_k)^2]$ | $x^3 f(x)^2$ | $x^3$ | $x$ | $x^2$ |
| $[\sum_k {p_k}^2 f(p_k)]^2$ | $[\sum_k x_k^2 f(x_k)]^2$ | $[\sum_k x_k^2]^2$ | $xy$ | $\sqrt{x} y^{\frac{3}{2}}$ |
| $\Delta E(Z^2)$ | | $\geq 0$ | $\geq 0$ | Indefinite |

## REFERENCES

[1] M.-C. BABRON, M. MARTINEZ, C. BONTAITI-PELLIE, AND F. CLERGET-DARPOUX, *Linkage detection by affected-pedigree-member method: what is really tested?*, Genet. Epidemiol., 10 (1993), pp. 389–394.

[2] L. R. GOLDIN AND D. E. WEEKS, *Two-locus models of diseases: Comparison of likelihood and nonparametric linkage methods*, Am. J. Hum. Genet., 53 (1993), pp. 908–915.

[3] M. SCHROEDER, D. L. BROWN, AND D. E. WEEKS, *Improved programs for the affected-pedigree-member method of linkage analysis*, Genet. Epidimiol., 11 (1994), pp. 69–74.

[4] P. VAN EERDEWEGH, C. L. HAMPE, B. K. SUAREZ, AND T. REICH, *Alzheimer's disease: a piscatorial trek*, Genet. Epidemiol., 10 (1993), pp. 395–400.

[5] D. E. WEEKS AND K. LANGE, *The affected-pedigree-member method of linkage analysis*, Am. J. Hum. Genet., 42 (1988), pp. 315–326.

# TRIANGLE CONSTRAINTS FOR SIB-PAIR IDENTITY BY DESCENT PROBABILITIES UNDER A GENERAL MULTILOCUS MODEL FOR DISEASE SUSCEPTIBILITY

SANDRINE DUDOIT* AND TERENCE P. SPEED*

**Abstract.** In this paper, we study sib-pair IBD probabilities under a general multilocus model for disease susceptibility which doesn't assume random mating, linkage equilibrium or Hardy-Weinberg equilibrium. We derive the triangle constraints satisfied by affected, discordant and unaffected sib-pair IBD probabilities, as well as constraints distinguishing between sharing of maternal and paternal DNA, under general monotonicity assumptions concerning the penetrance probabilities. The triangle constraints are valid for age and sex-dependent penetrances, and in the presence of parental imprinting. We study the parameterization of sib-pair IBD probabilities for common models, and present examples to demonstrate the impact of non-random mating and the necessity of our assumptions for the triangle constraints. We prove that the affected sib-pair possible triangle is covered by the IBD probabilities of two types of models, one with fixed mode of inheritance and general mating type frequencies, the other with varying mode of inheritance and random mating. Finally, we consider IBD probabilities at marker loci linked to disease susceptibility loci and derive the triangle constraints satisfied by these probabilities.

**Key words.** Identity by descent, linkage, disease genes, affected sib-pair method, triangle constraints, multilocus sib-pair IBD probabilities, Hardy-Weinberg equilibrium, random mating, linkage equilibrium, imprinting, age and sex-dependent penetrances.

**1. Introduction.** The affected sib-pair method is used routinely to test for linkage between a marker locus and a disease susceptibility (DS) locus. The method consists of sampling nuclear families with two affected children and establishing the number of chromosomes on which the two sibs share DNA identical by descent (IBD) at a marker. The observed distribution $(N_0, N_1, N_2)$ for the number of affected sib-pairs sharing DNA IBD at the marker on 0, 1, and 2 chromosomes, respectively, is then compared to the expected proportions under random Mendelian segregation, $(\frac{1}{4}, \frac{1}{2}, \frac{1}{4})$. Deviations from the $(\frac{1}{4}, \frac{1}{2}, \frac{1}{4})$ null distribution are taken as evidence of linkage between the marker and a DS locus. Several test statistics have been proposed to test for linkage, such as the "mean IBD" statistic, $N_2 + \frac{1}{2}N_1$ (Blackwelder and Elston [2], Knapp *et al.* [16, 17, 18]). In order to increase the power to detect DS loci, other tests have been suggested which make use of the triangle constraints satisfied by the affected-sib pair IBD probabilities: $\pi_1 \leq \frac{1}{2}$ and $2\pi_0 \leq \pi_1$, where $\pi_i$ is the probability that an affected sib-pair shares DNA IBD at a DS locus on $i$, $i = 0, 1, 2$, chromosomes (Holmans [13], Faraway [9], Holmans and Clayton [14], Cordell *et al.* [3], Knapp *et al.* [18]). For a single DS locus model, these constraints have only been proved under the stringent assumptions of random mating and Hardy-Weinberg equilibrium at the DS locus (Suarez *et al.* [35], Holmans

*Department of Statistics, University of California, Berkeley, CA 94720-3860.

[13]). For a model with two unlinked DS loci, the constraints were proved with the additional assumption of linkage equilibrium between the two loci (Cordell *et al.* [3]).

In this paper, we study sib-pair IBD probabilities under a general multilocus model for disease susceptibility which doesn't assume random mating, linkage equilibrium or Hardy-Weinberg equilibrium. We derive the triangle constraints satisfied by affected, discordant and unaffected sib-pair IBD probabilities, as well as constraints distinguishing between sharing of maternal and paternal DNA, under general monotonicity assumptions concerning the penetrance probabilities. The triangle constraints are valid for age and sex-dependent penetrances, and in the presence of parental imprinting. We study the parameterization of sib-pair IBD probabilities for common models, and present examples to demonstrate the impact of non-random mating and the necessity of our assumptions for the triangle constraints. We prove that the affected sib-pair possible triangle is covered by the IBD probabilities of two types of models, one with fixed mode of inheritance and general mating type frequencies, the other with varying mode of inheritance and random mating. In general, a triple $(\pi_0, \pi_1, \pi_2)$ satisfying the triangle constraints corresponds to the IBD probabilities for many different modes of inheritance, thus it is inappropriate to estimate the IBD probabilities and solve for parameters such as penetrances and allele frequencies, unless one has knowledge of the mode of inheritance. Finally, we consider IBD probabilities at marker loci linked to disease susceptibility loci and derive the triangle constraints satisfied by these probabilities. Note that we do not address the question of linkage testing in this paper, but derive properties of the IBD probabilities which may be used subsequently in devising appropriate tests of linkage. Although this paper is concerned with sib-pair IBD probabilities, we present the basic definitions, models and derivations for sibships of arbitrary size and phenotype pattern in order to retain the generality of our approach. We used this approach to derive score tests of linkage for general sibships and optimal weights for combining such test statistics across the various types of sibships [7].

The remainder of this section presents basic definitions and an overview of the affected sib-pair method. We introduce the general multilocus model in Section 2 and derive the conditional distribution of inheritance vectors at DS loci given the phenotype vector of a sibship in Section 3. In Sections 4 and 5, we derive the triangle constraints for a single DS locus model and for a general multilocus model. In Section 6, we consider a single diallelic DS locus and study the parameterization of sib-pair IBD probabilities. Finally, in Section 7, we consider IBD probabilities at marker loci linked to disease susceptibility loci.

**1.1. Identity by descent.** DNA at the same locus on two homologous chromosomes is said to be *identical by descent (IBD)* if it originated from the same ancestral chromosome. If two homologous chromosomes from *different* people are IBD at some locus, the people are *related*. If two homologous chromosomes from the *same* person are IBD at some locus, this person is *inbred*, i.e. has related parents. Two people, neither of whom is inbred, can share DNA IBD at a particular locus on either 0, 1, or 2 chromosomes.

**1.2. Inheritance vectors.** Consider a sibship of $k \geq 2$ sibs, and suppose we wish to identify the parental origin of the DNA inherited by each sib at a particular autosomal locus, $\mathcal{L}$ say. Arbitrarily label the paternal chromosomes containing the locus of interest by (1,2), and similarly label the maternal chromosomes by (3,4). The labeling of parental chromosomes is done independently for unlinked loci. The *inheritance vector* (also called *gene-identity state* or *vector of segregation indicators*) of the sibship at the locus $\mathcal{L}$ is the $2k$-vector

$$X(\mathcal{L}) = x = (x_1, x_2, ..., x_{2k-1}, x_{2k}),$$

where for $1 \leq i \leq k$,

$$\begin{aligned} x_{2i-1} &= \text{label of paternal chromosome from which} \\ &\quad\ \text{sib } i \text{ inherited DNA at } \mathcal{L} \\ &= 1 \text{ or } 2; \\ x_{2i} &= \text{label of maternal chromosome from which} \\ &\quad\ \text{sib } i \text{ inherited DNA at } \mathcal{L} \\ &= 3 \text{ or } 4. \end{aligned}$$

According to the above definition there are $2^{2k}$ inheritance vectors for the sibship. However, inheritance vectors obtained by permuting the labels 1 and 2 and/or 3 and 4 represent the same IBD configuration in terms of sharing of paternal and maternal DNA, so there are only $2^{2(k-1)}$ distinct inheritance vectors. There may be further collapsing of the inheritance vectors into *IBD configurations* in the case of affected only sibships (Ethier and Hodge [8]). Note that the labels 1, 2, 3 and 4 for the parental chromosomes are only meaningful within a sibship and may correspond to different DNA sequences in different sibships. Hence, these "alleles" are neither transportable across families nor functional. If there is no inbreeding at $\mathcal{L}$, then 1, 2, 3 and 4 represent DNA sequences that are distinct by descent. Here we only consider IBD sharing resulting from the $2k$ segregations giving rise to the sibship, and allow inbreeding in the population. In practice, the inheritance vector of a sibship is determined by finding enough polymorphism in the parents to be able to identify the chromosomal fragments transmitted to individuals in the sibship. When IBD information is incomplete, partial information extracted from marker data may be summarized by the *inheritance distribution*, a conditional probability distribution over the possible inheritance vectors at the marker locus (Kruglyak and Lander [20], Kruglyak *et al.* [19]). Risch [33], Holmans [13] and Holmans and Clayton [14] also address the issue of incomplete marker polymorphism.

For sib-pairs, one usually considers three distinct IBD configurations at a particular locus, according to the number of chromosomes sharing DNA IBD at the locus. In some cases (e.g. parental imprinting), it may be appropriate to distinguish the parental origin of the DNA shared IBD by the sib-pair, and consider four IBD configurations as shown in Table 1.

TABLE 1
*Sib-pair IBD configurations.*

| Number IBD | Representative inheritance vector |
|---|---|
| 0 | (1,3,2,4) |
| 1 paternal | (1,3,1,4) |
| 1 maternal | (1,3,2,3) |
| 2 | (1,3,1,3) |

**1.3. Phenotype vector.** For a disease of interest, denote the *phenotype vector* of the $k$ sibs by the vector $\phi$

$$\phi = (\phi_1, ..., \phi_k),$$

where for $1 \le i \le k$, $\phi_i$ is the phenotype indicator of the $i$th sib

$$\phi_i = \begin{cases} 1, & \text{if the } i\text{th sib is affected,} \\ 0, & \text{if the } i\text{th sib is unaffected.} \end{cases}$$

For sib-pairs, there are three phenotype patterns, corresponding to the number of affected sibs.

TABLE 2
*Sib-pair phenotype vectors.*

| Pattern (abbreviation) | Number of affected sibs | Phenotype vector |
|---|---|---|
| Affected sib-pair (ASP) | 2 | (1,1) |
| Discordant sib-pair (DSP) | 1 | (1,0) or (0,1) |
| Unaffected sib-pair (USP) | 0 | (0,0) |

**1.4. The affected sib-pair method.** In general, there is an *association* between phenotype and IBD configuration of related individuals at loci linked to DS loci. This may be illustrated in a simple case, by considering IBD sharing and phenotype in sib-pairs for a fully recessive DS locus $\mathcal{D}$, with alleles $D$ and $d$, where $D$ is recessive with respect to $d$ (i.e. only individuals with genotype $DD$ are affected). To simplify computation, we further assume that both parents are heterozygous. Considering all 16 possible transmission patterns from these two parents to the two children, we build up tables of joint probabilities of phenotype and IBD configuration at the DS locus and at a locus unlinked to the DS locus. Table 3 clearly indicates an *association* between phenotype and IBD configuration *at* the DS locus, while Table 4 indicates *independence* of phenotype and IBD configuration at a locus *unlinked* to the DS locus.

TABLE 3

*Joint probability of # affected sibs and # chromosomes sharing DNA IBD* at *DS locus.*

| | # affected sibs 0 | 1 | 2 | |
|---|---|---|---|---|
| # chromosomes 0 | $\frac{1}{8}$ | $\frac{1}{8}$ | 0 | $\frac{1}{4}$ |
| sharing DNA IBD 1 | $\frac{1}{4}$ | $\frac{1}{4}$ | 0 | $\frac{1}{2}$ |
| *at* DS locus 2 | $\frac{3}{16}$ | 0 | $\frac{1}{16}$ | $\frac{1}{4}$ |
| | $\frac{9}{16}$ | $\frac{3}{8}$ | $\frac{1}{16}$ | |

TABLE 4

*Joint probability of # affected sibs and # chromosomes sharing DNA IBD at a locus* unlinked *to the DS locus.*

| | # affected sibs 0 | 1 | 2 | |
|---|---|---|---|---|
| # chromosomes 0 | $\frac{9}{64}$ | $\frac{3}{32}$ | $\frac{1}{64}$ | $\frac{1}{4}$ |
| sharing DNA IBD at a 1 | $\frac{9}{32}$ | $\frac{3}{16}$ | $\frac{1}{32}$ | $\frac{1}{2}$ |
| locus *unlinked* to the DS locus 2 | $\frac{9}{64}$ | $\frac{3}{32}$ | $\frac{1}{64}$ | $\frac{1}{4}$ |
| | $\frac{9}{16}$ | $\frac{3}{8}$ | $\frac{1}{16}$ | |

This association suggests the following strategy for mapping disease genes: take groups of related individuals with particular disease phenotypes and examine the frequencies with which specific IBD configurations arise at candidate DS loci. The most popular strategy is the *affected sib-pair method*, which studies IBD sharing between two sibs affected with the disease of interest. In 1975, Cudworth and Woodrow [5] considered the IBD distribution of the HLA haplotypes of 15 sib-pairs affected with juvenile-onset diabetes mellitus and compared this distribution to the proportions of $(\frac{1}{4}, \frac{1}{2}, \frac{1}{4})$ expected under random Mendelian segregation. They found a significant deviation from the $(\frac{1}{4}, \frac{1}{2}, \frac{1}{4})$ distribution and their study initiated a large body of research on the implication of HLA and other loci in insulin-dependent diabetes mellitus (IDDM) (Day and Simons [6], Thomson and Bodmer [36], Suarez *et al.* [35], Motro and Thomson [26], Louis *et al.* [23, 22], Payami *et al.* [29, 28], Cox and Spielman [4], to name a few). Since then, the affected sib-pair method has been studied extensively, initially in the context of HLA-disease association and subsequently for various complex diseases (Alzheimer disease [30], schizophrenia [1], atopy [24]) and genome scans (Kruglyak and Lander [20], Feingold *et al.* [10], Feingold and Siegmund [11]). Day and Simons [6] derived the IBD distribution of affected sib-pairs and affected cousin-pairs at a single random mating diallelic DS locus for quasi-recessive and quasi-dominant modes of inheritance (see Section 6). Suarez *et al.* [35] derived the IBD distribution for the three types of sib-pairs (ASPs, DSPs and USPs) at a marker linked to a diallelic random mating DS locus. Motro and Thomson [26], Louis *et al.* [23, 22] and Payami *et al.* [29, 28] considered the problem of estimating the mode of inheritance and the allele frequency for a single diallelic DS locus with random mating, using IBD data from affected sib-pairs and affected sib-trios. Louis *et al.* [22] and Payami *et al.* [29] also studied the impact of selection, non-random mating, meiotic drive and recombination on the IBD probabilities. Risch [31, 32] considered multilocus models, with the usual assumptions of random mating and Hardy-Weinberg equilibrium at the DS loci and linkage equilibrium between the DS loci, and expressed the IBD probabilities of affected relative pairs in terms of $\lambda_R$, the risk ratio for relatives of type $R$ compared with population prevalence.

Several test statistics have been suggested to test the null hypothesis of no linkage including the "mean IBD" statistic $N_2 + \frac{1}{2}N_1$ (Blackwelder and Elston [2], Knapp *et al.* [16, 17, 18]), $N_2 + \frac{1}{4}N_1$ (Feingold and Siegmund [11]), likelihood ratio statistics and $\chi^2$ goodness-of-fit statistics, either unrestricted or restricted to the "possible triangle" (Risch [32, 33], Holmans [13], Faraway [9], Feingold *et al.* [10], Holmans and Clayton [14], Cordell *et al.* [3], Knapp *et al.* [18]).

Until now, the properties of the ASP method have been studied mainly under the population genetic assumptions of Hardy-Weinberg equilibrium, random mating, and linkage equilibrium for the DS loci. However, these population genetic assumptions are not only likely to be violated for most diseases of interest, but are hard to verify and their violation has a potentially large impact on the IBD probabilities (cf. Section 6.4). In what follows, we will be concerned with deriving the conditional probabilities of inheritance vectors given the phenotype vector of a sibship, as well as inequalities satisfied by these probabilities, under the general genetic model introduced in the next section.

## 2. Genetic model.

The genetic model consists of three main components: a model for disease susceptibility, connecting disease phenotypes to genotypes at DS loci in the groups of related individuals of interest (Section 2.1); a population genetic model, describing the population joint distribution of genotypes at the DS loci for the relevant founders (Section 2.2); and a segregation model, describing the segregation of alleles at the DS loci during meiosis (Section 2.3).

### 2.1. Model for disease susceptibility.

#### 2.1.1. Basic model.

In our general model, we consider $L$ unlinked autosomal DS loci, $\mathcal{D}^1, \ldots, \mathcal{D}^L$, where $\mathcal{D}^l$ has $m_l$ alleles, $D_1^l, \ldots, D_{m_l}^l$, $l = 1, \ldots, L$. We define the *multilocus penetrance* of a genotype at the $L$ DS loci to be the conditional probability of affectedness given the multilocus genotype at the $L$ DS loci, i.e.

$$f_{i_1 j_1, \ldots, i_L j_L} = pr(\text{Affected} \mid D_{i_1}^1 D_{j_1}^1, \ldots, D_{i_L}^L D_{j_L}^L),$$

where $i_l,\ j_l = 1, \ldots, m_l,\ \ l = 1, \ldots, L$, and

$$\begin{aligned} D_{i_l}^l &= \text{paternally inherited allele at locus } \mathcal{D}^l, \\ D_{j_l}^l &= \text{maternally inherited allele at locus } \mathcal{D}^l. \end{aligned}$$

This definition allows the possibility of *parental imprinting*, i.e. different paternal and maternal contributions to disease susceptibility, as observed with Prader-Willi and Angelman syndromes (Lalande [21], Niikawa [27]). The involvement of imprinting was also suggested in the aetiology of atopy (Moffatt *et al.* [24]), IDDM and bipolar affective disorder (Lalande [21]). Special cases of these general penetrances include the multiplicative, additive and heterogeneity models of Risch [31].

In order to derive constraints satisfied by conditional IBD probabilities, we make the following assumption about the dependence structure of genotypes and phenotypes within a family:

▷ `Assumption G1.` Within a family, the phenotype of a particular sib is conditionally independent of the phenotypes and genotypes of his siblings, given his multilocus genotype at $\mathcal{D}^1, \ldots, \mathcal{D}^L$. That is, for a family of $k$ sibs

$$pr(\phi_1, \ldots, \phi_k | sg_1, \ldots, sg_k) = \prod_{i=1}^{k} pr(\phi_i | sg_i),$$

where $\phi_i$ and $sg_i$ denote the phenotype and multilocus genotype of the $i$th sib, respectively. This assumption rules out environmental covariance in the sib phenotypes.

Note that the *marginal* penetrances for genotypes at single DS loci depend on conditional genotype frequencies. For example, for $\mathcal{D}^1$

$$pr(\text{Affected} \mid D^1_{i_1} D^1_{j_1}) \\ = \sum_{i_2 j_2, \dots, i_L j_L} f_{i_1 j_1, \dots, i_L j_L}\, pr(D^2_{i_2} D^2_{j_2}, \dots, D^L_{i_L} D^L_{j_L} \mid D^1_{i_1} D^1_{j_1}).$$

Also, we have assumed conditional independence of sib phenotypes given *multilocus* genotypes at *all* DS loci, and not given *marginal* genotypes at *individual* loci. Hence, even when computing conditional IBD probabilities at a single DS locus, we need to condition on the genotypes at all DS loci.

**2.1.2. Age and sex-dependent penetrances.** We may also consider a more general model that allows age and sex-dependent penetrances as follows:

$$f^a_{i_1 j_1, \dots, i_L j_L} = pr(\text{Affected} \mid D^1_{i_1} D^1_{j_1}, \dots, D^L_{i_L} D^L_{j_L}, a),$$

where $a$ denotes the age and sex of a particular individual. Autosomal dominant inheritance with age-dependent penetrances has been used to explain the familial aggregation of Alzheimer disease (Pericak-Vance *et al.* [30]), and sex-dependent penetrances may be involved in IDDM (Morahan *et al.* [25]). `Assumption G1` then becomes

▷ `Assumption G1b`. Within a family, the phenotype of a particular sib is conditionally independent of any phenotype, genotype, age and sex data on his siblings, given his multilocus genotype at $\mathcal{D}^1, \dots, \mathcal{D}^L$, and his age and sex. That is, for a family of $k$ sibs

$$pr(\phi_1, \dots, \phi_k | sg_1, \dots, sg_k, a_1, \dots, a_k) = \prod_{i=1}^{k} pr(\phi_i | sg_i, a_i),$$

where $\phi_i$, $sg_i$ and $a_i$ denote the phenotype, multilocus genotype, age and sex of the $i$th sib, respectively. We can extend this model to accommodate other types of covariates.

**2.2. Population genetic model.** In order to derive the conditional distribution of inheritance vectors given the phenotype vector of a sibship, we will need to refer to the pairs of genotypes possessed by the parents at the DS loci. Let $pg^l = (pg^l_1, pg^l_2, pg^l_3, pg^l_4)$ denote the *ordered parental genotype* at the DS locus $\mathcal{D}^l$, $l = 1, \dots, L$, where $pg^l_i$ is the allele at $\mathcal{D}^l$ on the parental chromosome labeled $i$, $i = 1, 2, 3, 4$. For a DS locus with $m$ alleles, there are $m^2 \times m^2$ ordered parental genotypes. These may be grouped into $(m(m+1)/2)^2$ *parental mating types*, by grouping genotypes which may be obtained from one another by permuting alleles 1 and 2 and/or 3 and 4. Let $mt^l$ denote the parental mating type at the DS locus $\mathcal{D}^l$, and let $pg = (pg^1, \dots, pg^L)$ and $mt = (mt^1, \dots, mt^L)$ denote the multilocus ordered parental genotypes and mating types, respectively (see Table 5). For unlinked DS loci, because of the independent labeling of parental chromosomes, all ordered genotypes within a mating type have the same frequency. Hence

$$pr(pg^1, \ldots, pg^L) = \frac{pr(mt^1, \ldots, mt^L)}{\prod_{l=1}^{L} \#\{pg \in mt^l\}},$$

where $\#\{pg \in mt^l\}$ is the number of ordered parental genotypes which are part of the mating type $mt^l$.

TABLE 5
*Representative parental mating type and ordered genotypes at $\mathcal{D}^l$.*

| Parental mating type $mt^l$ | Parental genotypes $pg^l$ |
|---|---|
| $[hijk]$ | $D_h^l D_i^l \times D_j^l D_k^l$<br>$D_h^l D_i^l \times D_k^l D_j^l$<br>$D_i^l D_h^l \times D_j^l D_k^l$<br>$D_i^l D_h^l \times D_k^l D_j^l$ |

Most authors assume Hardy-Weinberg equilibrium and random mating at the DS loci, as well as linkage equilibrium between the loci. These assumptions would give expressions for the mating type frequencies in terms of a series $\{p_i^l\}$ of allele frequencies at each DS locus $\mathcal{D}^l$. We avoid these problematic assumptions as far as possible, and study the impact of non-random mating on the IBD probabilities in Section 6.4. Our general model also allows for the possibility of inbreeding at the DS loci and selective disadvantage on affected individuals (Louis *et al.* [22] and Payami *et al.* [29]). A lot may be gained in generality by dropping the usual population genetic assumptions, and with minimal cost in terms of added complexity.

**2.3. Segregation model.** We will make one last assumption in order to derive constraints satisfied by conditional probabilities of inheritance vectors. Let $x^l$ denote the inheritance vector of a sibship of size $k$ at the DS locus $\mathcal{D}^l$, $l = 1, \ldots, L$, and $x = (x^1, \ldots, x^L)$.

▷ `Assumption G2.` There is no *segregation distortion*, i.e. for a sibship of size $k$ with multilocus parental genotype $pg$ and multilocus inheritance vector $x$ at unlinked DS loci

$$pr(x|pg) = \left(\frac{1}{4^k}\right)^L.$$

## 3. Multilocus conditional distribution of inheritance vectors at DS loci given phenotype vector of sibship.

**3.1. Basic model.** We will compute the multilocus conditional distribution $pr(x|\phi)$ of the inheritance vectors at all $L$ DS loci given the phenotype vector $\phi$ of a sibship of size $k$ under our general model. By Bayes Theorem and under `Assumption G2`:

$$(1) \qquad pr(x|\phi) = \frac{\sum_{pg} pr(\phi|x, pg)\, pr(x|pg)\, pr(pg)}{\sum_x \sum_{pg} pr(\phi|x, pg)\, pr(x|pg)\, pr(pg)} = \frac{\sum_{pg} pr(\phi|x, pg)\, pr(pg)}{\sum_x \sum_{pg} pr(\phi|x, pg)\, pr(pg)}.$$

Now, $x$ and $pg$ together yield the multilocus ordered sib genotypes (distinguishing paternally and maternally inherited alleles), $sg_1, \ldots, sg_k$. Hence, under `Assumption G1`:

$$pr(x|\phi) = \frac{\sum_{pg^1} \cdots \sum_{pg^L} \{\prod_{i=1}^k pr(\phi_i|sg_i)\}\, pr(pg^1, \ldots, pg^L)}{\sum_{x^1} \cdots \sum_{x^L} \sum_{pg^1} \cdots \sum_{pg^L} \{\prod_{i=1}^k pr(\phi_i|sg_i)\}\, pr(pg^1, \ldots, pg^L)},$$

where $1 \le i \le k$ labels individual sibs and

$$pr(\phi_i|sg_i) = pr(\phi_i \mid pg^l_{x^l_{2i-1}}\, pg^l_{x^l_{2i}},\ l = 1, \ldots, L).$$

This multilocus conditional distribution is a function of the multilocus penetrances and the mating type frequencies which we denote by the global parameter $\nu$.

The following proposition allows us to derive constraints satisfied by the IBD probabilities regardless of the mating type frequencies, by deriving a sufficient condition for the constraints which doesn't involve the population genetic model.

PROPOSITION 1. *Under* `Assumption G2`*, a sufficient condition for the following inequality*

$$(2) \qquad \forall\, x^2, \ldots, x^L \quad \sum_{x^1} c(x^1)\, pr(x^1, \ldots, x^L|\phi) \ge 0$$

*is* $\forall\, mt^1$, $\forall\, pg^2, \ldots, pg^L$, $\forall\, x^2, \ldots, x^L$

$$(3) \qquad \sum_{x^1} c(x^1) \left\{ \sum_{\{pg^1 \in mt^1\}} pr(\phi|x^1, \ldots, x^L, pg^1, \ldots, pg^L) \right\} \ge 0.$$

*Proof.* For all $x^2, \ldots, x^L$, equation (3) implies

$$\sum_{pg^2} \cdots \sum_{pg^L} \sum_{mt^1} \frac{pr(mt^1, \ldots, mt^L)}{\prod_{l=1}^L \#\{pg \in mt^l\}} \times \sum_{x^1} c(x^1) \left\{ \sum_{\{pg^1 \in mt^1\}} pr(\phi|x^1, \ldots, x^L, pg^1, \ldots, pg^L) \right\} \ge 0$$

$$\Rightarrow \sum_{x^1} c(x^1) \sum_{pg} pr(\phi|x^1,\ldots,x^L,pg^1,\ldots,pg^L)\, pr(pg^1,\ldots,pg^L) \geq 0$$

$$\Rightarrow \sum_{x^1} c(x^1)\, \frac{\sum_{pg} pr(\phi|x^1,\ldots,x^L,pg)\, pr(pg)}{\sum_x \sum_{pg} pr(\phi|x^1,\ldots,x^L,pg)\, pr(pg)} \geq 0$$

$$\Rightarrow \sum_{x^1} c(x^1)\, pr(x^1,\ldots,x^L|\phi) \geq 0, \quad \text{which is (2).} \qquad \square$$

This generalizes in the obvious way for constraints at other loci.

**3.2. Age and sex-dependent penetrances.** In the presence of age and sex-dependent penetrances (or penetrances depending on other covariates), two approaches are possible. In the first, we stratify the IBD data according to phenotype, age and sex, and compute the conditional distribution of the inheritance vectors given the phenotype vector $\phi$ and the age and sex information $a$ of the sibship. Then, by **Assumptions G1b, G2**:

$$pr(x|\phi,a) = \frac{\sum_{pg} pr(\phi|x,pg,a)\, pr(x|pg,a)\, pr(pg|a)}{\sum_x \sum_{pg} pr(\phi|x,pg,a)\, pr(x|pg,a)\, pr(pg|a)}$$

$$= \frac{\sum_{pg} \{\prod_{i=1}^k pr(\phi_i|sg_i,a_i)\}\, pr(pg)}{\sum_x \sum_{pg} \{\prod_{i=1}^k pr(\phi_i|sg_i,a_i)\}\, pr(pg)}.$$

With the second approach, we compute the conditional distribution of the inheritance vectors given only $\phi$ as follows:

$$pr(x|\phi) = \frac{\sum_{pg} \sum_a pr(\phi|x,pg,a)\, pr(x|pg,a)\, pr(pg|a)\, pr(a)}{\sum_x \sum_{pg} \sum_a pr(\phi|x,pg,a)\, pr(x|pg,a)\, pr(pg|a)\, pr(a)}$$

$$= \frac{\sum_{pg} \sum_a \{\prod_{i=1}^k pr(\phi_i|sg_i,a_i)\}\, pr(pg)\, pr(a)}{\sum_x \sum_{pg} \sum_a \{\prod_{i=1}^k pr(\phi_i|sg_i,a_i)\}\, pr(pg)\, pr(a)},$$

where $pr(a)$ is the hypothetical probability of a sibship with age and sex $a$. $pr(x|\phi)$ may be expressed as a mixture of the IBD probabilities conditional on the age and sex information of the sibship,

$$pr(x|\phi) = \sum_a pr(a|\phi)\; pr(x|\phi,a).$$

Consequently, if there are large differences in penetrances between ages and/or sexes (Morahan *et al.* [25]), stratification as in the first approach may increase power to detect linkage in one of the age/sex groups. In either case, Proposition 1 may be modified to accommodate age and sex-dependent penetrances as follows:

PROPOSITION 2. *Under* **Assumption G2** *and for a model with age and sex-dependent penetrances, a sufficient condition for the following inequalities*

$$\forall\, x^2,\ldots,x^L \quad \sum_{x^1} c(x^1)\, pr(x^1,\ldots,x^L|\phi) \geq 0$$

*and*

$$\forall\, a,\ \forall\, x^2,\ldots,x^L \quad \sum_{x^1} c(x^1)\, pr(x^1,\ldots,x^L|\phi,a) \geq 0$$

*is* $\forall\, a,\ \forall\, mt^1,\ \forall\, pg^2,\ldots,pg^L,\ \forall\, x^2,\ldots,x^L$

$$\sum_{x^1} c(x^1) \left\{ \sum_{\{pg^1 \in mt^1\}} pr(\phi|x^1,\ldots,x^L,pg^1,\ldots,pg^L,a) \right\} \geq 0\,.$$

Hence, the age and sex data $a$ on a sibship may be treated as the parental genotype at another unlinked DS locus, at least for the purpose of deriving constraints. In order to highlight the main ideas in our approach for deriving the triangle constraints, without the complexity of notation introduced by multilocus models with age and sex-dependent penetrances, we will defer the treatment of these models to Section 5. Unless specified otherwise, we will consider the basic model of Section 2.1.1.

**4. Constraints for sib-pair conditional IBD probabilities under a general single autosomal disease locus model.** In this section we consider a single autosomal DS locus $\mathcal{D}$, with $m$ alleles, $D_1,\ldots,D_m$, and arbitrary mating type frequencies, and we are concerned with deriving constraints satisfied by sib-pair conditional IBD probabilities at this locus.

For affected sib-pairs, let

$$\begin{aligned}\pi_i^{ASP}(\nu) &= pr(\text{Sib-pair shares DNA IBD on } i \text{ chromosomes at } \mathcal{D}|ASP)\\ &= \sum_{C_i} pr(x|ASP),\end{aligned}$$

where $C_i = \{x : \delta(x_1,x_3)+\delta(x_2,x_4) = i\}$, $i = 0,1,2$, $x = (x_1,x_2,x_3,x_4)$ denotes the inheritance vector of the sib-pair at $\mathcal{D}$, and $\delta(k,l) = 1$ if $k = l$ and 0, otherwise. In some cases (e.g. parental imprinting), we may be interested in distinguishing between sharing of maternal and paternal DNA by the sib-pair, so let

$$\begin{aligned}\pi_{ij}^{ASP}(\nu) &= pr(\text{Sib-pair shares DNA IBD at } \mathcal{D} \text{ on } i \text{ paternal and}\\ &\qquad j \text{ maternal chromosomes}|ASP)\\ &= \sum_{C_{ij}} pr(x|ASP),\end{aligned}$$

where $C_{ij} = \{x : \delta(x_1,x_3) = i,\ \delta(x_2,x_4) = j\}$, $i,\ j = 0,1$. Then

$$\pi_{10}^{ASP}(\nu) + \pi_{01}^{ASP}(\nu) = \pi_1^{ASP}(\nu).$$

The DSP and USP IBD probabilities are defined similarly, and we may drop the parameter $\nu$ and the sib-pair type to simplify notation when there is no ambiguity. Examples of models for which $\pi_{01} \neq \pi_{10}$ are given in Section 6. We will prove constraints satisfied by the sib-pair conditional IBD probabilities under our general single DS locus model and the following monotonicity assumption concerning the penetrances:

▷ `Assumption M1.` $\forall\, i, j, k, l = 1, \ldots, m$

[M1a] $$(f_{ik} - f_{jk})(f_{il} - f_{jl}) \geq 0,$$

and

[M1b] $$(f_{ik} - f_{il})(f_{jk} - f_{jl}) \geq 0.$$

For symmetric penetrances (i.e. no parental imprinting), this is equivalent to the existence of an ordering of the alleles at the DS locus such that:

$$\forall\, i = 1, \ldots, m \qquad f_{i1} \leq f_{i2} \leq \ldots \leq f_{im}.$$

`Assumption M1` is satisfied by the usual diallelic recessive, dominant and additive modes of inheritance, but not by over-dominant modes of inheritance (e.g. $f_{11}, f_{22} < f_{12} = f_{21}$). It is also satisfied in the case of parental imprinting where paternally and maternally inherited alleles are ordered differently in terms of "severity" (e.g. $f_{21} \leq f_{11} \leq f_{12}$, $f_{21} \leq f_{22} \leq f_{12}$, where $D_2$ is protective if paternally inherited, but increases susceptibility if maternally inherited).

PROPOSITION 3. *Under a general single autosomal DS locus model with m alleles, arbitrary mating type frequencies, and* `Assumptions G1, G2, M1`, *the ASP conditional IBD probabilities at the DS locus satisfy the following constraints:*

$$\pi_{10}^{ASP}(\nu) + \pi_{01}^{ASP}(\nu) \leq \pi_{00}^{ASP}(\nu) + \pi_{11}^{ASP}(\nu), \tag{4}$$

$$\pi_{00}^{ASP}(\nu) \leq \pi_{10}^{ASP}(\nu), \tag{5}$$

$$\pi_{00}^{ASP}(\nu) \leq \pi_{01}^{ASP}(\nu). \tag{6}$$

*Consequently,*

$$\pi_{1}^{ASP}(\nu) \leq \frac{1}{2}, \tag{7}$$

$$\pi_{1}^{ASP}(\nu) \geq 2\pi_{0}^{ASP}(\nu). \tag{8}$$

*Hence, the ASP IBD probabilities at the DS locus fall in a triangle with vertices*

$$\left(\frac{1}{4}, \frac{1}{2}, \frac{1}{4}\right), \quad \left(0, \frac{1}{2}, \frac{1}{2}\right) \quad \text{and} \quad (0, 0, 1)$$

*which we call the ASP possible triangle. The USP IBD probabilities satisfy the same constraints.*

Note that the IBD probabilities satisfy the possible triangle constraints under the assumptions of random mating and Hardy-Weinberg equilibrium at the DS locus, without requiring `Assumption M1`. However, in real mapping situations, it is more likely to encounter non-random mating than modes of inheritance which violate `Assumption M1`. The ASP possible triangle is also referred to in the literature as Holmans' possible triangle.

*Proof.* The proof relies on Proposition 1 and the fact that if $\tilde{x}$ is obtained from $x$ by permuting 1 and 2 and/or 3 and 4, then $\forall\ mt$

$$\sum_{\{pg \in mt\}} pr(\phi|\tilde{x}, pg) = \sum_{\{pg \in mt\}} pr(\phi|x, pg). \tag{9}$$

TABLE 6

*Conditional probability of ASP given inheritance vector $x$ and parental genotype $pg$ for a representative mating type $mt$.*

$pr(ASP|x, pg)$

| Parental genotype | Inheritance vector $x$ | | | |
|---|---|---|---|---|
| $pg$ | (1,3,1,3) | (1,3,1,4) | (1,3,2,3) | (1,3,2,4) |
| $D_iD_j \times D_kD_l$ | $f_{ik}^2$ | $f_{ik}f_{il}$ | $f_{ik}f_{jk}$ | $f_{ik}f_{jl}$ |
| $D_iD_j \times D_lD_k$ | $f_{il}^2$ | $f_{il}f_{ik}$ | $f_{il}f_{jl}$ | $f_{il}f_{jk}$ |
| $D_jD_i \times D_kD_l$ | $f_{jk}^2$ | $f_{jk}f_{jl}$ | $f_{jk}f_{ik}$ | $f_{jk}f_{il}$ |
| $D_jD_i \times D_lD_k$ | $f_{jl}^2$ | $f_{jl}f_{jk}$ | $f_{jl}f_{il}$ | $f_{jl}f_{ik}$ |

- To prove that $\pi_{10}^{ASP}(\nu)+\pi_{01}^{ASP}(\nu) \leq \pi_{00}^{ASP}(\nu)+\pi_{11}^{ASP}(\nu)$, it suffices to show that $\forall\ mt$

$$\sum_{\{pg \in mt\}} pr(ASP|(1,3,1,4), pg) + \sum_{\{pg \in mt\}} pr(ASP|(1,3,2,3), pg)$$
$$\leq \sum_{\{pg \in mt\}} pr(ASP|(1,3,2,4), pg) + \sum_{\{pg \in mt\}} pr(ASP|(1,3,1,3), pg).$$

This inequality is true since

$$\sum_{\{pg \in mt\}} pr(ASP|(1,3,2,4), pg) + \sum_{\{pg \in mt\}} pr(ASP|(1,3,1,3), pg)$$
$$- \sum_{\{pg \in mt\}} pr(ASP|(1,3,1,4), pg) - \sum_{\{pg \in mt\}} pr(ASP|(1,3,2,3), pg)$$
$$= f_{ik}(f_{ik} + f_{jl} - f_{il} - f_{jk}) + f_{il}(f_{il} + f_{jk} - f_{ik} - f_{jl})$$
$$+ f_{jk}(f_{jk} + f_{il} - f_{jl} - f_{ik}) + f_{jl}(f_{jl} + f_{ik} - f_{jk} - f_{il})$$
$$= (f_{ik} + f_{jl} - f_{il} - f_{jk})^2 \geq 0.$$

- To prove that $\pi_{10}^{ASP}(\nu) \geq \pi_{00}^{ASP}(\nu)$, it suffices to show that $\forall\ mt$

$$\sum_{\{pg \in mt\}} pr(ASP|(1,3,1,4), pg) \geq \sum_{\{pg \in mt\}} pr(ASP|(1,3,2,4), pg).$$

This inequality is true since

$$\begin{aligned}
&\sum_{\{pg \in mt\}} pr(ASP|(1,3,1,4), pg) - \sum_{\{pg \in mt\}} pr(ASP|(1,3,2,4), pg) \\
&= f_{ik}(f_{il} - f_{jl}) + f_{il}(f_{ik} - f_{jk}) + f_{jk}(f_{jl} - f_{il}) + f_{jl}(f_{jk} - f_{ik}) \\
&= (f_{ik} - f_{jk})(f_{il} - f_{jl}) + (f_{il} - f_{jl})(f_{ik} - f_{jk}) \\
&= 2(f_{ik} - f_{jk})(f_{il} - f_{jl}) \geq 0 \text{ under } \texttt{Assumption M1a}.
\end{aligned}$$

- The proof of $\pi_{01}^{ASP}(\nu) \geq \pi_{00}^{ASP}(\nu)$ is similar and involves

$$\begin{aligned}
&\sum_{\{pg \in mt\}} pr(ASP|(1,3,2,3), pg) - \sum_{\{pg \in mt\}} pr(ASP|(1,3,2,4), pg) \\
&= 2(f_{ik} - f_{il})(f_{jk} - f_{jl}) \geq 0 \text{ under } \texttt{Assumption M1b}.
\end{aligned}$$

Equation (8) follows immediately from (5) and (6). Equation (4) implies $\pi_1 \leq \pi_0 + \pi_2 = 1 - \pi_1$, which is (7).

The proof for USPs is similar to that for ASPs, but with $f_{ij}$ replaced by its complement $1 - f_{ij}$. □

PROPOSITION 4. *Under a general single autosomal DS locus model with m alleles, arbitrary mating type frequencies, and* Assumptions G1, G2, M1, *the DSP conditional IBD probabilities at the DS locus satisfy the following constraints:*

$$\pi_{00}^{DSP}(\nu) + \pi_{11}^{DSP}(\nu) \leq \pi_{10}^{DSP}(\nu) + \pi_{01}^{DSP}(\nu), \tag{10}$$

$$\pi_{10}^{DSP}(\nu) \leq \pi_{00}^{DSP}(\nu), \tag{11}$$

$$\pi_{01}^{DSP}(\nu) \leq \pi_{00}^{DSP}(\nu). \tag{12}$$

*Consequently,*

$$\pi_1^{DSP}(\nu) \geq \frac{1}{2}, \tag{13}$$

$$\pi_1^{DSP}(\nu) \leq 2\pi_0^{DSP}(\nu). \tag{14}$$

*Hence, the DSP IBD probabilities at the DS locus fall in a triangle with vertices*

$$\left(\frac{1}{4}, \frac{1}{2}, \frac{1}{4}\right), \quad \left(\frac{1}{2}, \frac{1}{2}, 0\right) \text{ and } \left(\frac{1}{3}, \frac{2}{3}, 0\right)$$

*which we call the DSP possible triangle.*

*Proof.* The proof relies on Proposition 1, equation (9) and the fact that $\forall\ x$ and $\forall\ mt$

$$\sum_{\{pg\in mt\}} pr((1,0)|x,pg) = \sum_{\{pg\in mt\}} pr((0,1)|x,pg).$$

TABLE 7

*Conditional probability of DSP given inheritance vector $x$ and parental genotype $pg$ for a representative mating type $mt$.*

$$pr((1,0)|x,pg)$$

| Parental genotype | Inheritance vector $x$ | | | |
|---|---|---|---|---|
| $pg$ | (1,3,1,3) | (1,3,1,4) | (1,3,2,3) | (1,3,2,4) |
| $D_iD_j \times D_kD_l$ | $f_{ik}(1-f_{ik})$ | $f_{ik}(1-f_{il})$ | $f_{ik}(1-f_{jk})$ | $f_{ik}(1-f_{jl})$ |
| $D_iD_j \times D_lD_k$ | $f_{il}(1-f_{il})$ | $f_{il}(1-f_{ik})$ | $f_{il}(1-f_{jl})$ | $f_{il}(1-f_{jk})$ |
| $D_jD_i \times D_kD_l$ | $f_{jk}(1-f_{jk})$ | $f_{jk}(1-f_{jl})$ | $f_{jk}(1-f_{ik})$ | $f_{jk}(1-f_{il})$ |
| $D_jD_i \times D_lD_k$ | $f_{jl}(1-f_{jl})$ | $f_{jl}(1-f_{jk})$ | $f_{jl}(1-f_{il})$ | $f_{jl}(1-f_{ik})$ |

The rest of the proof is similar to the proof for ASPs, and involves the following quantities:

- $$\sum_{\{pg\in mt\}} pr((1,0)|(1,3,1,4),pg) + \sum_{\{pg\in mt\}} pr((1,0)|(1,3,2,3),pg) - \sum_{\{pg\in mt\}} pr((1,0)|(1,3,2,4),pg) - \sum_{\{pg\in mt\}} pr((1,0)|(1,3,1,3),pg) = (f_{ik}+f_{jl}-f_{il}-f_{jk})^2 \geq 0.$$

- $$\sum_{\{pg\in mt\}} pr((1,0)|(1,3,2,4),pg) - \sum_{\{pg\in mt\}} pr((1,0)|(1,3,1,4),pg) = 2(f_{ik}-f_{jk})(f_{il}-f_{jl}) \geq 0$$ under `Assumption M1a`.

- $$\sum_{\{pg\in mt\}} pr((1,0)|(1,3,2,4),pg) - \sum_{\{pg\in mt\}} pr((1,0)|(1,3,2,3),pg) = 2(f_{ik}-f_{il})(f_{jk}-f_{jl}) \geq 0$$ under `Assumption M1b`. □

Since the trinomial probabilities $(\pi_0,\pi_1,\pi_2)$ must be nonnegative and add up to unity, the triple $(\pi_0,\pi_1,\pi_2)$ corresponds to a point in the simplex

$$\mathcal{S} = \{(\pi_0,\pi_1,\pi_2) : \pi_i \geq 0,\ i=0,1,2,\ \text{and}\ \pi_0+\pi_1+\pi_2 = 1\}.$$

A convenient way of displaying the trinomial probabilities is using a *barycentric representation.* Barycentric coordinates in the plane represent the triple $(\pi_0, \pi_1, \pi_2)$ by the vector $\pi_0 A_0 + \pi_1 A_1 + \pi_2 A_2$, where the $A_i$'s are fixed vectors in the plane, such as the columns of the $2 \times 3$ matrix in the following equation:

$$\begin{bmatrix} x \\ y \end{bmatrix} = \begin{bmatrix} \sqrt{2} & \frac{\sqrt{2}}{2} & 0 \\ 0 & \sqrt{\frac{3}{2}} & 0 \end{bmatrix} \times \begin{bmatrix} \pi_0 \\ \pi_1 \\ \pi_2 \end{bmatrix}.$$

With this representation, $(0,0,1)$ is located at the origin and $(\pi_0, \pi_1, \pi_2)$ are points in an equilateral triangle with sides of length $\sqrt{2}$. The vertices of the triangle correspond to one of the $\pi$'s being unity, and along the sides of the triangle one of the $\pi$'s is zero (see Figures 1, 2 p. 198). Holmans [13] uses a different representation for the trinomial probabilities which is two-dimensional and involves only $(\pi_0, \pi_1)$. The boundaries of the space for $(\pi_0, \pi_1)$ are $\pi_0 = 0$, $\pi_1 = 0$ and $\pi_1 + \pi_0 = 1$, and the boundaries of the ASP possible triangle are $\pi_0 = 0$, $\pi_1 = \frac{1}{2}$ and $\pi_1 = 2\pi_0$.

**5. Constraints for sib-pair conditional IBD probabilities under a general multilocus model.** In this section we consider a general model with $L$ unlinked autosomal DS loci and define sib-pair multilocus IBD probabilities as follows. For $i_l, j_l = 0, 1, \ l = 1, \ldots, L$, let

$$\begin{aligned}\pi^{ASP}_{i_1 j_1, \ldots, i_L j_L}(\nu) &= pr(\text{Sib-pair shares DNA IBD at } \mathcal{D}^l \text{ on } i_l \text{ paternal} \\ &\qquad \text{and } j_l \text{ maternal chromosomes}, l = 1, \ldots, L \mid ASP) \\ &= \sum_{\mathcal{C}_{i_1 j_1, \ldots, i_L j_L}} pr(x^1, \ldots, x^L | ASP),\end{aligned}$$

where

$$\mathcal{C}_{i_1 j_1, \ldots, i_L j_L} = \{(x^1, \ldots, x^L) : \ \delta(x_1^l, x_3^l) = i_l, \ \delta(x_2^l, x_4^l) = j_l, \ l = 1, \ldots, L\},$$

and $x^l$ is the inheritance vector of the sib-pair at the DS locus $\mathcal{D}^l$. The marginal IBD probabilities at the DS locus $\mathcal{D}^l$, $l = 1, \ldots, L$, are defined by

$$\begin{aligned}\pi^{ASP}_{++, \ldots, i_l j_l, \ldots, ++}(\nu) &= pr(\text{Sib-pair shares DNA IBD at } \mathcal{D}^l \text{ on } i_l \\ &\qquad \text{paternal and } j_l \text{ maternal chromosomes} \mid ASP) \\ &= \sum_{\mathcal{C}_{++, \ldots, i_l j_l, \ldots, ++}} pr(x^1, \ldots, x^L | ASP),\end{aligned}$$

where $\mathcal{C}_{++, \ldots, i_l j_l, \ldots, ++} = \{(x^1, \ldots, x^L) : \ \delta(x_1^l, x_3^l) = i_l, \ \delta(x_2^l, x_4^l) = j_l\}$. DSP and USP IBD probabilities are defined similarly.

We will derive constraints satisfied by the sib-pair multilocus IBD probabilities under our general multilocus model and the following monotonicity assumption which is a generalization of `Assumption M1` for multiple loci:

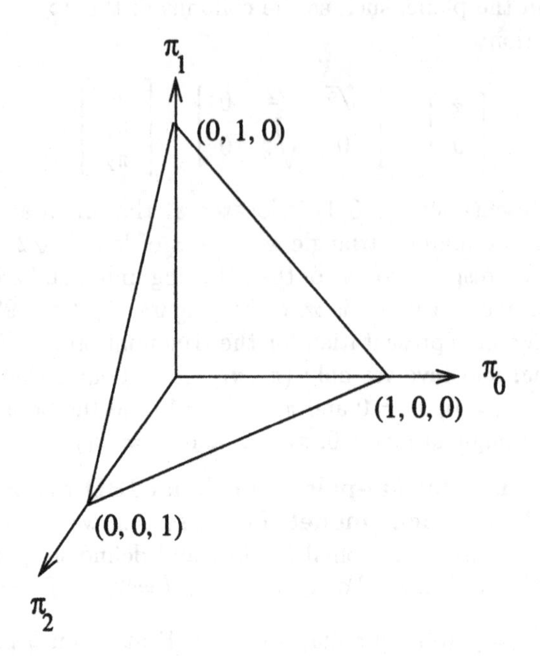


FIG. 1. *Simplex $\mathcal{S}$.*

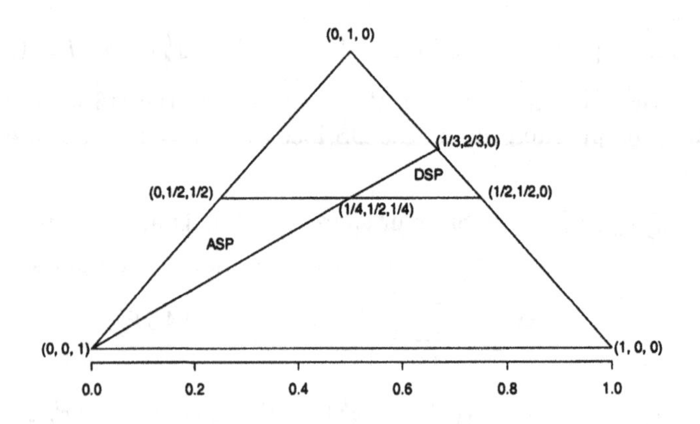


FIG. 2. *ASP and DSP possible triangles.*

▷ `Assumption M2`. For each DS locus $\mathcal{D}^h$, $h = 1, \ldots, L$, let $g$, $\tilde{g}$ denote any two multilocus ordered genotypes at the remaining $L-1$ loci. Then $\forall\, h = 1, \ldots, L, \forall\, g,\ \tilde{g}$ and $\forall\, i,\ j,\ k,\ l = 1, \ldots, m_h$

[M2a] $$(f_{ik,g} - f_{jk,g})(f_{il,\tilde{g}} - f_{jl,\tilde{g}}) \geq 0,$$

[M2b] $$(f_{ik,g} - f_{il,g})(f_{jk,\tilde{g}} - f_{jl,\tilde{g}}) \geq 0,$$

[M2c] $$(f_{ik,g} + f_{jl,g} - f_{il,g} - f_{jk,g})(f_{ik,\tilde{g}} + f_{jl,\tilde{g}} - f_{il,\tilde{g}} - f_{jk,\tilde{g}}) \geq 0,$$

where $f_{ik,g}$ is the probability of affectedness given ordered genotype $D_i^h D_k^h$ at $\mathcal{D}^h$ and $g$ at the remaining $L-1$ loci.

This assumption is satisfied by Risch's [31] multiplicative and additive penetrances, with the single locus constraints (`Assumption M1`) holding at each DS locus.

PROPOSITION 5. *Under a general multilocus model with arbitrary mating type frequencies and* `Assumptions G1,G2, M2`, *the ASP (and USP) multilocus IBD probabilities at the DS loci satisfy the following constraints. For each DS locus $\mathcal{D}^l$, $l=1, \ldots, L$, and $\forall\, i_h,\ j_h = 0, 1,\ h = 1, \ldots, L,\ h \neq l$*

$$\pi^{ASP}_{i_1 j_1, \ldots, 00, \ldots, i_L j_L}(\nu) \leq \pi^{ASP}_{i_1 j_1, \ldots, 10, \ldots, i_L j_L}(\nu), \tag{15}$$

$$\pi^{ASP}_{i_1 j_1, \ldots, 00, \ldots, i_L j_L}(\nu) \leq \pi^{ASP}_{i_1 j_1, \ldots, 01, \ldots, i_L j_L}(\nu), \tag{16}$$

$$\begin{aligned}\pi^{ASP}_{i_1 j_1, \ldots, 01, \ldots, i_L j_L}(\nu) &+ \pi^{ASP}_{i_1 j_1, \ldots, 10, \ldots, i_L j_L}(\nu) \\ &\leq \pi^{ASP}_{i_1 j_1, \ldots, 00, \ldots, i_L j_L}(\nu) + \pi^{ASP}_{i_1 j_1, \ldots, 11, \ldots, i_L j_L}(\nu).\end{aligned} \tag{17}$$

*Consequently, for each DS locus, the marginal IBD probabilities satisfy*

$$\pi^{ASP}_{++, \ldots, 00, \ldots, ++}(\nu) \leq \pi^{ASP}_{++, \ldots, 10, \ldots, ++}(\nu), \tag{18}$$

$$\pi^{ASP}_{++, \ldots, 00, \ldots, ++}(\nu) \leq \pi^{ASP}_{++, \ldots, 01, \ldots, ++}(\nu), \tag{19}$$

$$\begin{aligned}\pi^{ASP}_{++, \ldots, 01, \ldots, ++}(\nu) &+ \pi^{ASP}_{++, \ldots, 10, \ldots, ++}(\nu) \\ &\leq \pi^{ASP}_{++, \ldots, 00, \ldots, ++}(\nu) + \pi^{ASP}_{++, \ldots, 11, \ldots, ++}(\nu),\end{aligned} \tag{20}$$

*and hence the single locus ASP possible triangle constraints are satisfied. The DSP multilocus IBD probabilities satisfy the reverse inequalities*

$$\pi^{DSP}_{i_1 j_1, \ldots, 00, \ldots, i_L j_L}(\nu) \geq \pi^{DSP}_{i_1 j_1, \ldots, 10, \ldots, i_L j_L}(\nu), \tag{21}$$

$$\pi^{DSP}_{i_1 j_1, \ldots, 00, \ldots, i_L j_L}(\nu) \geq \pi^{DSP}_{i_1 j_1, \ldots, 01, \ldots, i_L j_L}(\nu), \tag{22}$$

$$\begin{aligned}\pi^{DSP}_{i_1 j_1, \ldots, 01, \ldots, i_L j_L}(\nu) &+ \pi^{DSP}_{i_1 j_1, \ldots, 10, \ldots, i_L j_L}(\nu) \\ &\geq \pi^{DSP}_{i_1 j_1, \ldots, 00, \ldots, i_L j_L}(\nu) + \pi^{DSP}_{i_1 j_1, \ldots, 11, \ldots, i_L j_L}(\nu).\end{aligned} \tag{23}$$

*It follows that the marginal DSP IBD probabilities at each DS locus fall in the single locus DSP possible triangle.*

*Proof.* Without loss of generality, we will prove the inequalities for $\mathcal{D}^1$ using Proposition 1. We will use the fact that if $\tilde{x}^1$ is obtained from $x^1$ by permuting the labels 1 and 2 and/or 3 and 4, then

$$\sum_{\{pg^1 \in mt^1\}} pr(\phi|\tilde{x}^1,\ldots,x^L,pg^1,\ldots,pg^L) = \sum_{\{pg^1 \in mt^1\}} pr(\phi|x^1,\ldots,x^L,pg^1,\ldots,pg^L).$$

TABLE 8

*Conditional probability of ASP given inheritance vectors $x^1,\ldots,x^L$ and parental genotypes $pg^1,\ldots,pg^L$ for a representative mating type $mt^1$. $sg_1$ and $sg_2$ denote the multilocus genotypes of the two sibs at the last $L-1$ loci, as specified by $x^2,\ldots,x^L$ and $pg^2,\ldots,pg^L$.*

$$pr(ASP|x^1,\ldots,x^L,pg^1,\ldots,pg^L)$$

| $pg^1$ | Inheritance vector at DS locus $\mathcal{D}^1$, $x^1$ | | | |
|---|---|---|---|---|
| | (1,3,1,3) | (1,3,1,4) | (1,3,2,3) | (1,3,2,4) |
| $D_i^1D_j^1 \times D_k^1D_l^1$ | $f_{ik,sg_1}f_{ik,sg_2}$ | $f_{ik,sg_1}f_{il,sg_2}$ | $f_{ik,sg_1}f_{jk,sg_2}$ | $f_{ik,sg_1}f_{jl,sg_2}$ |
| $D_i^1D_j^1 \times D_l^1D_k^1$ | $f_{il,sg_1}f_{il,sg_2}$ | $f_{il,sg_1}f_{ik,sg_2}$ | $f_{il,sg_1}f_{jl,sg_2}$ | $f_{il,sg_1}f_{jk,sg_2}$ |
| $D_j^1D_i^1 \times D_k^1D_l^1$ | $f_{jk,sg_1}f_{jk,sg_2}$ | $f_{jk,sg_1}f_{jl,sg_2}$ | $f_{jk,sg_1}f_{ik,sg_2}$ | $f_{jk,sg_1}f_{il,sg_2}$ |
| $D_j^1D_i^1 \times D_l^1D_k^1$ | $f_{jl,sg_1}f_{jl,sg_2}$ | $f_{jl,sg_1}f_{jk,sg_2}$ | $f_{jl,sg_1}f_{il,sg_2}$ | $f_{jl,sg_1}f_{ik,sg_2}$ |

• To prove that $\pi^{ASP}_{00,i_2j_2,\ldots,i_Lj_L}(\nu) \leq \pi^{ASP}_{10,i_2j_2,\ldots,i_Lj_L}(\nu)$, it suffices to show that $\forall\ mt^1$, $\forall\ pg^2,\ldots,pg^L$, $\forall\ x^2,\ldots,x^L$

$$\sum_{\{pg^1 \in mt^1\}} pr(ASP|(1,3,1,4),x^2,\ldots,x^L,pg^1,\ldots,pg^L) \geq \sum_{\{pg^1 \in mt^1\}} pr(ASP|(1,3,2,4),x^2,\ldots,x^L,pg^1,\ldots,pg^L).$$

This inequality is true since under `Assumption M2a`

$$\begin{aligned}
&\sum_{\{pg^1 \in mt^1\}} pr(ASP|(1,3,1,4),x^2,\ldots,x^L,pg^1,\ldots,pg^L) \\
&\qquad - \sum_{\{pg^1 \in mt^1\}} pr(ASP|(1,3,2,4),x^2,\ldots,x^L,pg^1,\ldots,pg^L) \\
&= f_{ik,sg_1}(f_{il,sg_2} - f_{jl,sg_2}) + f_{il,sg_1}(f_{ik,sg_2} - f_{jk,sg_2}) \\
&\qquad + f_{jk,sg_1}(f_{jl,sg_2} - f_{il,sg_2}) + f_{jl,sg_1}(f_{jk,sg_2} - f_{ik,sg_2}) \\
&= (f_{ik,sg_1} - f_{jk,sg_1})(f_{il,sg_2} - f_{jl,sg_2}) \\
&\qquad + (f_{il,sg_1} - f_{jl,sg_1})(f_{ik,sg_2} - f_{jk,sg_2}) \geq 0.
\end{aligned}$$

• The proof of $\pi^{ASP}_{00,i_2j_2,\ldots,i_Lj_L}(\nu) \leq \pi^{ASP}_{01,i_2j_2,\ldots,i_Lj_L}(\nu)$ involves expressions similar to those above, and `Assumption M2b`.

• To prove that

$$\pi^{ASP}_{01,i_2j_2,\ldots,i_Lj_L}(\nu)+\pi^{ASP}_{10,i_2j_2,\ldots,i_Lj_L}(\nu)\leq\pi^{ASP}_{00,i_2j_2,\ldots,i_Lj_L}(\nu)+\pi^{ASP}_{11,i_2j_2,\ldots,i_Lj_L}(\nu)$$

it suffices to show that $\forall\ mt^1,\ \forall\ pg^2,\ldots,pg^L,\ \forall\ x^2,\ldots,x^L$

$$\begin{aligned}
&\sum_{\{pg^1\in mt^1\}} pr(ASP|(1,3,1,4),x^2,\ldots,x^L,pg^1,\ldots,pg^L)\\
&\quad+\sum_{\{pg^1\in mt^1\}} pr(ASP|(1,3,2,3),x^2,\ldots,x^L,pg^1,\ldots,pg^L)\\
\leq&\sum_{\{pg^1\in mt^1\}} pr(ASP|(1,3,1,3),x^2,\ldots,x^L,pg^1,\ldots,pg^L)\\
&\quad+\sum_{\{pg^1\in mt^1\}} pr(ASP|(1,3,2,4),x^2,\ldots,x^L,pg^1,\ldots,pg^L).
\end{aligned}$$

This inequality is true since under `Assumption M2c`

$$\begin{aligned}
&\sum_{\{pg^1\in mt^1\}} pr(ASP|(1,3,1,3),x^2,\ldots,x^L,pg^1,\ldots,pg^L)\\
&\quad+\sum_{\{pg^1\in mt^1\}} pr(ASP|(1,3,2,4),x^2,\ldots,x^L,pg^1,\ldots,pg^L)\\
&\quad-\sum_{\{pg^1\in mt^1\}} pr(ASP|(1,3,1,4),x^2,\ldots,x^L,pg^1,\ldots,pg^L)\\
&\quad-\sum_{\{pg^1\in mt^1\}} pr(ASP|(1,3,2,3),x^2,\ldots,x^L,pg^1,\ldots,pg^L)\\
=\ &f_{ik,sg_1}(f_{ik,sg_2}+f_{jl,sg_2}-f_{il,sg_2}-f_{jk,sg_2})\\
&\quad+f_{il,sg_1}(f_{il,sg_2}+f_{jk,sg_2}-f_{ik,sg_2}-f_{jl,sg_2})\\
&\quad+f_{jk,sg_1}(f_{jk,sg_2}+f_{il,sg_2}-f_{jl,sg_2}-f_{ik,sg_2})\\
&\quad+f_{jl,sg_1}(f_{jl,sg_2}+f_{ik,sg_2}-f_{jk,sg_2}-f_{il,sg_2})\\
=\ &(f_{ik,sg_1}+f_{jl,sg_1}-f_{il,sg_1}-f_{jk,sg_1})\\
&\cdot(f_{ik,sg_2}+f_{jl,sg_2}-f_{il,sg_2}-f_{jk,sg_2})\geq 0.
\end{aligned}$$

The proof of the USP inequalities follows immediately by replacing the penetrances by their complements. The inequalities for DSPs also follow immediately by replacing the penetrances involving $sg_2$ by their complements. □

The proof of corresponding constraints for age and sex-dependent penetrances, with or without conditioning on the age and sex information of the sibship, is similar and relies on Proposition 2. It involves penetrances of the form $f^a_{ik,g}$ and constraints similar to those in `Assumption M2`, namely: $\forall\ a,\tilde{a},\ \forall\ h=1,\ldots,L,\forall\ g,\ \tilde{g}$ and $\forall\ i,\ j,\ k,\ l=1,\ldots,m_h$

$$(f^a_{ik,g} - f^a_{jk,g})(f^{\bar{a}}_{il,\bar{g}} - f^{\bar{a}}_{jl,\bar{g}}) \geq 0,$$
$$(f^a_{ik,g} - f^a_{il,g})(f^{\bar{a}}_{jk,\bar{g}} - f^{\bar{a}}_{jl,\bar{g}}) \geq 0,$$
$$(f^a_{ik,g} + f^a_{jl,g} - f^a_{il,g} - f^a_{jk,g})(f^{\bar{a}}_{ik,\bar{g}} + f^{\bar{a}}_{jl,\bar{g}} - f^{\bar{a}}_{il,\bar{g}} - f^{\bar{a}}_{jk,\bar{g}}) \geq 0.$$

Under Risch's [31] multiplicative model with Hardy-Weinberg equilibrium, random mating and linkage equilibrium,

$$pr(x^1, \ldots, x^L|\phi) = \prod_{l=1}^{L} pr(x^l|\phi),$$

consequently,

$$\pi_{i_1 j_1, \ldots, i_L j_L} = \prod_{l=1}^{L} \pi_{++, \ldots, i_l j_l, \ldots, ++}.$$

**6. Single diallelic DS locus models.** In this section, we will study the parameterization of the ASP IBD probabilities under common models for disease susceptibility. We will give examples of genetic models for which the triangle constraints are violated and examine the impact of non-random mating on the IBD probabilities. Note that the models considered in this section may not always be realistic, but are nevertheless useful for our purpose.

Consider a single DS locus $\mathcal{D}$ with two alleles: a "disease" allele, $D$, and a "wild-type" allele, $d$, and define three penetrance probabilities as follows:

$$f_2 = pr(\text{Affected} \mid DD),$$
$$f_1 = pr(\text{Affected} \mid Dd) = pr(\text{Affected} \mid dD),$$
$$f_0 = pr(\text{Affected} \mid dd).$$

Common models for the penetrances are:

- *Strict-recessive:* $0 = f_0 = f_1 < f_2$ (Thomson and Bodmer [36]);
- *Quasi-recessive:* $0 < f_0 = f_1 \leq f_2 = rf_0$ (Day and Simons [6]);
- *Strict-dominant:* $0 = f_0 < f_1 = f_2$ (Thomson and Bodmer [36]);
- *Quasi-dominant:* $0 < f_0 \leq f_1 = f_2 = rf_0$ (Day and Simons [6]);
- *Additive:* $f_1 = \frac{f_0+f_2}{2}$ (Motro and Thomson [26]);
- *Intermediate:* $0 = f_0 \leq f_1 = sf_2 \leq f_2$ (Spielman *et al.* [34], Louis *et al.* [23]).

There are nine different mating types (Table 9), with frequencies denoted by $P_{mt}$, $mt = 1, \ldots, 9$.

**6.1. Random mating and Hardy-Weinberg equilibrium models.** Denote the frequencies of alleles $D$ and $d$ in the population of interest by $p$ and $q = 1 - p$, respectively. Assume that mating is random at $\mathcal{D}$

TABLE 9

*Parental mating types, ordered genotypes and their frequencies for a diallelic DS locus with random mating and Hardy-Weinberg equilibrium.*

| Parental mating type, $mt$ | Parental genotypes, $pg$ | Random mating and HW frequencies, $P_{mt}$ |
|---|---|---|
| 1 | $DD \times DD$ | $p^4$ |
| 2 | $DD \times Dd$ | $p^3q$ |
| | $DD \times dD$ | $p^3q$ |
| 3 | $Dd \times DD$ | $p^3q$ |
| | $dD \times DD$ | $p^3q$ |
| 4 | $DD \times dd$ | $p^2q^2$ |
| 5 | $dd \times DD$ | $p^2q^2$ |
| 6 | $Dd \times Dd$ | $p^2q^2$ |
| | $Dd \times dD$ | $p^2q^2$ |
| | $dD \times Dd$ | $p^2q^2$ |
| | $dD \times dD$ | $p^2q^2$ |
| 7 | $Dd \times dd$ | $pq^3$ |
| | $dD \times dd$ | $pq^3$ |
| 8 | $dd \times Dd$ | $pq^3$ |
| | $dd \times dD$ | $pq^3$ |
| 9 | $dd \times dd$ | $q^4$ |

and the three genotypes $DD$, $Dd$ and $dd$ have the Hardy-Weinberg (HW) frequencies $p^2$, $2p(1-p)$ and $(1-p)^2$, respectively. We will give a detailed treatment of ASP IBD probabilities only, since they are more frequently used and involve simpler expressions than DSP and USP IBD probabilities. The ASP IBD probabilities at the DS locus only depend on the allele frequency $p$ and on the ratios of penetrances, $f_0/f_2$ and $f_1/f_2$. We will examine the ASP IBD probabilities under six common penetrance models, assuming random mating and Hardy-Weinberg equilibrium.

**6.1.1. Strict-recessive model.** (See Figure 3 p. 205.)

$$(\pi_0, \pi_1, \pi_2) = \left(\frac{p^2}{(1+p)^2}, \frac{2p}{(1+p)^2}, \frac{1}{(1+p)^2}\right), \quad 0 < p \leq 1.$$

The curve traced by the recessive probabilities when $p$ varies is the Hardy-Weinberg curve $\pi_1^2 = 4\pi_0\pi_2$ joining (0,0,1) to $(\frac{1}{4}, \frac{1}{2}, \frac{1}{4})$. Note that these IBD probabilities are independent of $f_2$. More generally, it may be shown that the ASP IBD probabilities lie on the Hardy-Weinberg curve if $f_1^2 = f_0 f_2$ (Knapp *et al.* [16]).

**6.1.2. Quasi-recessive model.** (See Figures 3 and 5 pp. 205, 207.)

$$\pi_0 = \frac{p^4r^2 + (2p^2(1-p^2))r + (1-p^2)^2}{p^2(1+p)^2r^2 + 2p^2(1-p)(3+p)r + (1-p)(-p^3-3p^2+4p+4)},$$

$$\pi_1 = \frac{2(1-2p^2+p^3+2p^2(1-p)r+p^3r^2)}{p^2(1+p)^2r^2+2p^2(1-p)(3+p)r+(1-p)(-p^3-3p^2+4p+4)},$$

$$\pi_2 = \frac{p^2r^2+1-p^2}{p^2(1+p)^2r^2+2p^2(1-p)(3+p)r+(1-p)(-p^3-3p^2+4p+4)}.$$

Note that $f_0$ cancels out, and as $r \to \infty$ we get the probabilities for the strictly recessive case. We proved that for fixed $p \in (0,1)$, the IBD probabilities lie on a line going from $(\frac{1}{4}, \frac{1}{2}, \frac{1}{4})$ to $(\frac{p^2}{(1+p)^2}, \frac{2p}{(1+p)^2}, \frac{1}{(1+p)^2})$, and given by

$$\pi_0 = \frac{1}{3+p}\left((1+p)-(1+3p)\pi_2\right),$$

$$\pi_1 = \frac{2}{3+p}\left(1+(p-1)\pi_2\right).$$

Hence, the trinomial probabilities may be re-parameterized as

$$t \longrightarrow t\left(\frac{1}{4}, \frac{1}{2}, \frac{1}{4}\right) + (1-t)\left(\frac{p^2}{(1+p)^2}, \frac{2p}{(1+p)^2}, \frac{1}{(1+p)^2}\right), \quad 0 < t \leq 1,$$

where $t \to 0$ yields the strict-recessive case, and $t = 1$ corresponds to the case of no allele influencing DS at the candidate locus. The parameter $t$ is used to obtain a simpler parameterization and has no direct genetic interpretation. $t$ may be expressed as a function of $(r, p)$

$$t = \frac{4(1-2p^2+2p^2r)}{(4-7p^2+2p^3+p^4)+2p^2(3-2p-p^2)r+p^2(1+p)^2r^2}.$$

Each point strictly between the Hardy-Weinberg curve and the line $\pi_1 = 2\pi_0$ ($\pi_1^2 < 4\pi_0\pi_2$ and $\pi_1 > 2\pi_0$) corresponds to the ASP conditional IBD probabilities for a unique (up to $f_0$) quasi-recessive model. The parameters of this model are

$$p = \frac{1-3\pi_0-\pi_2}{3\pi_2+\pi_0-1} \in (0,1), \tag{24}$$

$$r = \frac{4\pi_2(-2\pi_2^2+\pi_2(3-8\pi_0)+(-1+5\pi_0-6\pi_0^2))}{(-1+3\pi_0+\pi_2)(\pi_0^2+(-1+\pi_2)^2-2\pi_0(1+\pi_2))} + \frac{2(-1+\pi_0+3\pi_2)^2\sqrt{\pi_0}\sqrt{-1+2(\pi_0+\pi_2)}}{(-1+3\pi_0+\pi_2)(\pi_0^2+(-1+\pi_2)^2-2\pi_0(1+\pi_2))} \in (1,\infty). \tag{25}$$

**6.1.3. Additive model.** (See Figure 3 p. 205.)

$$\pi_0 = \frac{(f_0-f_0p+f_2p)^2}{4f_0^2-7f_0^2p+6f_0f_2p+f_2^2p+3f_0^2p^2-6f_0f_2p^2+3f_2^2p^2},$$

$$\pi_1 = \frac{1}{2},$$

$$\pi_2 = \frac{2f_0^2-3f_0^2p+2f_0f_2p+f_2^2p+f_0^2p^2-2f_0f_2p^2+f_2^2p^2}{2(4f_0^2-7f_0^2p+6f_0f_2p+f_2^2p+3f_0^2p^2-6f_0f_2p^2+3f_2^2p^2)}.$$

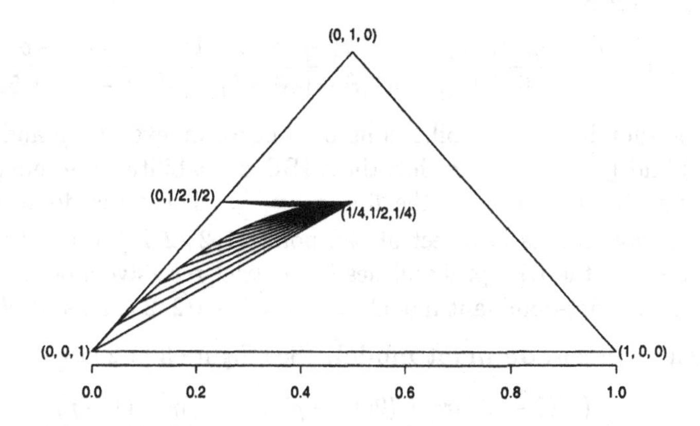


FIG. 3. *ASP quasi-recessive and quasi-dominant IBD probabilities. Strict-recessive model: the IBD probabilities lie on the Hardy-Weinberg curve joining (0,0,1) to* $(\frac{1}{4}, \frac{1}{2}, \frac{1}{4})$. *Quasi-recessive model: the lines under the Hardy-Weinberg curve are the IBD probabilities for fixed p and varying r. For fixed p, as r increases from 1 to* $\infty$, *the IBD probabilities move along a line from* $(\frac{1}{4}, \frac{1}{2}, \frac{1}{4})$ *to a point on the Hardy-Weinberg curve. Strict-dominant model: the IBD probabilities are on the curve joining* $(0, \frac{1}{2}, \frac{1}{2})$ *to* $(\frac{1}{4}, \frac{1}{2}, \frac{1}{4})$. *Quasi-dominant model: the lines above the strict-dominant curve are the IBD probabilities for fixed p and varying r.*

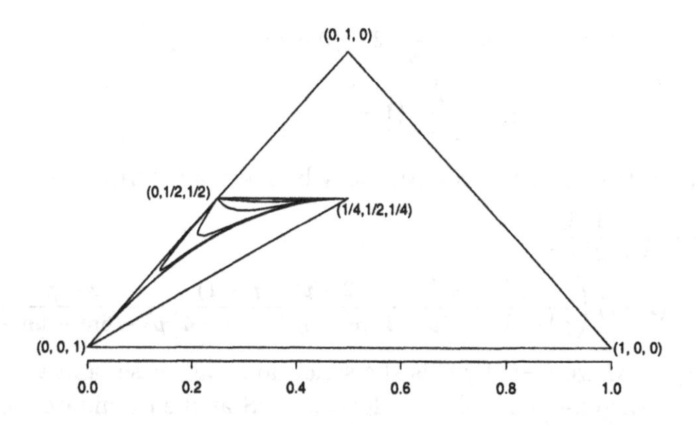


FIG. 4. *ASP intermediate IBD probabilities. From bottom to top, the curves are for fixed* $s = 0.01, 0.05, 0.1, 0.2, 0.3, 0.4$ *and varying* $0 < p < 1$.

These probabilities lie on the line $\pi_1 = \frac{1}{2}$ joining the points $(0, \frac{1}{2}, \frac{1}{2})$ and $(\frac{1}{4}, \frac{1}{2}, \frac{1}{4})$.

**6.1.4. Strict-dominant model.** (See Figure 3 p. 205.)

For $0 < p \leq 1$,

$$(\pi_0, \pi_1, \pi_2) = \left(\frac{p(2-p)^2}{p^3 - 6p^2 + 5p + 4}, \frac{2(-p^2 + p + 1)}{p^3 - 6p^2 + 5p + 4}, \frac{2-p}{p^3 - 6p^2 + 5p + 4}\right).$$

The strict-dominant probabilities lie on a curve indexed by $p$ and joining $(0, \frac{1}{2}, \frac{1}{2})$ and $(\frac{1}{4}, \frac{1}{2}, \frac{1}{4})$. Note that these IBD probabilities are independent of $f_2$ and are very close to the line $\pi_1 = \frac{1}{2}$. The strict-dominant and strict-recessive curves intersect at the point $(1, 2\sqrt{2}, 2)/(1 + \sqrt{2})^2$, which corresponds to the IBD probabilities for a strict-recessive model with $p = \sqrt{2}/2$ and a strict-dominant model with $p = 1 - \sqrt{2}/2$ (Louis *et al.* [23]).

**6.1.5. Quasi-dominant model.** (See Figure 3 p. 205.)

$$\pi_0 = \frac{(p^2(2-p)^2)r^2 + (2p(1-p)^2(2-p))r + (1-p)^4}{Den},$$
$$\pi_1 = \frac{2p(1+p-p^2)r^2 + 4p(1-2p+p^2)r + 2(1-p)^3}{Den},$$
$$\pi_2 = \frac{p(2-p)r^2 + (1-p)^2}{Den},$$

where

$$Den = p(4+5p-6p^2+p^3)r^2 + 2p(4-9p+6p^2-p^3)r + 4 - 12p + 13p^2 - 6p^3 + p^4.$$

Again, $f_0$ cancels out, and as $r \to \infty$ we get the probabilities for the strictly dominant case. We proved that for fixed $p \in (0, 1)$, the IBD probabilities lie on a line going through $(\frac{1}{4}, \frac{1}{2}, \frac{1}{4})$, and given by

$$\pi_0 = \frac{1}{4-p}((2-p) + (3p-4)\pi_2),$$
$$\pi_1 = \frac{2}{4-p}(1 - p\pi_2).$$

Hence, the trinomial probabilities may be re-parameterized as

$$t \longrightarrow t\left(\frac{1}{4}, \frac{1}{2}, \frac{1}{4}\right) + (1-t)\left(\frac{p(2-p)^2}{p^3 - 6p^2 + 5p + 4}, \frac{2(-p^2 + p + 1)}{p^3 - 6p^2 + 5p + 4}, \frac{2-p}{p^3 - 6p^2 + 5p + 4}\right),$$

$0 < t \leq 1$, where $t \to 0$ yields the strict-dominant case, and $t = 1$ corresponds to the case of no allele influencing DS at the candidate locus. Quasi-dominant probabilities are very close to the additive probabilities, i.e. to the line $\pi_1 = \frac{1}{2}$. Also, there is a small overlap between the IBD probabilities of quasi-recessive and quasi-dominant models, and a large region of the ASP triangle is not covered by either model.

**6.1.6. Intermediate model.** (See Figures 4 and 5 pp. 205, 207.)

$$\pi_0 = \frac{p\left(-2s + p\left(-1+2s\right)\right)^2}{p + p^3\left(1-2s\right)^2 + 4ps + 4s^2 + p^2\left(2-8s^2\right)},$$
$$\pi_1 = \frac{2\left(p^2\left(1-2s\right) - p\left(-2+s\right)s + s^2\right)}{p + p^3\left(1-2s\right)^2 + 4ps + 4s^2 + p^2\left(2-8s^2\right)},$$
$$\pi_2 = \frac{p + 2s^2 - 2ps^2}{p + p^3\left(1-2s\right)^2 + 4ps + 4s^2 + p^2\left(2-8s^2\right)}.$$

Special cases of the intermediate model include:

- $s = 0$: strict-recessive model;
- $s = \frac{1}{2}$: additive model with $f_0 = 0$

$$(\pi_0, \pi_1, \pi_2) = \left(\frac{p}{1+3p}, \frac{1}{2}, \frac{1+p}{2+6p}\right);$$

- $s = 1$: strict-dominant model.

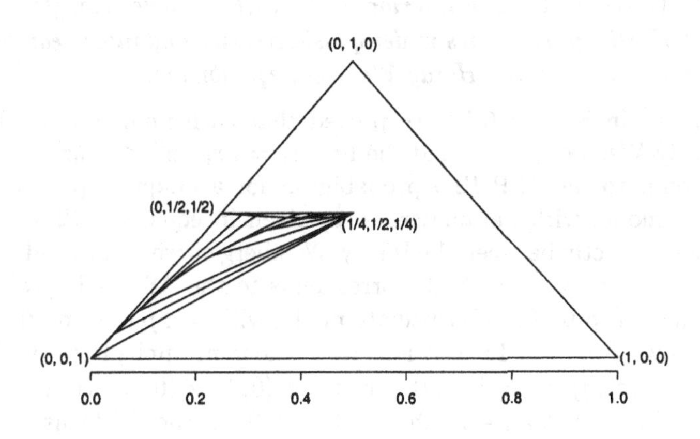


FIG. 5. *ASP quasi-recessive and intermediate IBD probabilities. Quasi-recessive model: as in Figure 3. Intermediate model: the curves joining the Hardy-Weinberg curve to the additive line are the IBD probabilities for intermediate models with fixed p and* $0 \le s \le \frac{1}{2}$*, while the curves joining the additive line to the strict-dominant curve are the IBD probabilities for intermediate models with fixed p and* $\frac{1}{2} \le s \le 1$*. From left to right* $p = 0.05, 0.1, 0.3, 0.5$*. These two models cover the interior of the ASP possible triangle.*

For fixed $p$, as $s$ increases from 0 to $\frac{1}{2}$, the IBD probabilities trace a curve from the Hardy-Weinberg curve to the line $\pi_1 = \frac{1}{2}$, and as $s$ increases from

$\frac{1}{2}$ to 1, the IBD probabilities trace a curve from the line $\pi_1 = \frac{1}{2}$ to the strict-dominant curve. For fixed $s \neq 0$, as $p$ increases from 0 to 1, the IBD probabilities trace a curve from $(0, \frac{1}{2}, \frac{1}{2})$ to $(\frac{1}{4}, \frac{1}{2}, \frac{1}{4})$.

The formulae for DSP and USP IBD probabilities are more complex since they involve one more parameter than the ASP probabilities. However, we proved that for a fixed disease allele frequency $p$, $0 < p < 1$, the quasi-recessive IBD probabilities for ASPs, DSPs and USPs fall on the *same line* going through $(\frac{1}{4}, \frac{1}{2}, \frac{1}{4})$. Similarly for the quasi-dominant probabilities. For an additive model, $\pi_1 = \frac{1}{2}$, and this also holds with arbitrary mating type frequencies, but in the latter case we may have $\pi_{10} \neq \pi_{01}$.

**6.2. Coverage of the ASP possible triangle.** The affected sib-pair possible triangle is covered by the IBD probabilities of two types of models, one with fixed mode of inheritance and general mating type frequencies, the other with varying mode of inheritance and random mating. In general, a point in the possible triangle corresponds to the IBD probabilities for many different modes of inheritance, thus it is inappropriate to estimate the IBD probabilities and solve for parameters such as penetrances and allele frequencies, unless one has knowledge of the mode of inheritance.

**6.2.1. Random mating and Hardy-Weinberg equilibrium models.**

PROPOSITION 6. *The interior of the ASP possible triangle is covered by the ASP IBD probabilities under quasi-recessive and intermediate models with random mating and Hardy-Weinberg equilibrium.*

*Proof.* In Section 6.1.2 we proved that each point strictly between the Hardy-Weinberg curve and the line $\pi_1 = 2\pi_0$ ($\pi_1^2 < 4\pi_0\pi_2$, $\pi_1 > 2\pi_0$) corresponds to the ASP IBD probabilities for a unique (up to $f_0$) quasi-recessive model with parameters $(r, p)$ given in equations (24) and (25). Each point strictly between the Hardy-Weinberg curve and the additive line ($\pi_1^2 > 4\pi_0\pi_2$, $\pi_1 < \frac{1}{2}$, $\pi_0 > 0$) corresponds to the ASP IBD probabilities for a unique (up to $f_2$) intermediate model with $0 < p < 1$ and $0 < s < \frac{1}{2}$. This follows from the Inverse Function Theorem applied to the function $(\pi_0(s,p), \pi_1(s,p))$ defined on the open set $(0, \frac{1}{2}) \times (0, 1)$. Here, $\pi_i(s,p)$ is the probability that an ASP shares DNA IBD at the DS locus on $i = 0, 1$ chromosomes under a random mating intermediate model with parameters $(s, p)$ and fixed $f_2$. The Jacobian matrix for this function is

$$J(s,p) = \begin{bmatrix} \frac{\partial \pi_0}{\partial s} & \frac{\partial \pi_0}{\partial p} \\ \frac{\partial \pi_1}{\partial s} & \frac{\partial \pi_1}{\partial p} \end{bmatrix}$$

and the determinant of the Jacobian matrix is

$$\begin{aligned} |J(s,p)| &= \frac{8\,(-1+p)^2\, p\, s\, (-1+2\,s)\, (-p-2\,s+2\,p\,s)\, (-p-s+2\,p\,s)}{\left(p + p^3\,(1-2\,s)^2 + 4\,p\,s + 4\,s^2 + p^2\,(2-8\,s^2)\right)^3} \\ &\neq 0 \qquad \text{for } (s,p) \in (0, \tfrac{1}{2}) \times (0, 1). \end{aligned}$$

Numerical solutions for $(s, p)$ may be obtained by the Newton-Raphson method (Louis *et al.* [23]). Finally, the Hardy-Weinberg curve is covered by ASP strict-recessive IBD probabilities. □

**6.2.2. Strict-recessive model.**

PROPOSITION 7. *The ASP possible triangle is covered by the ASP IBD probabilities for a strict-recessive model with general mating type frequencies. That is, for each* $(\pi_0, \pi_1, \pi_2)$ *with* $\pi_1 \leq \frac{1}{2}$, $\pi_0 \geq 0$ *and* $\pi_1 \geq 2\pi_0$, *we can find a family of mating type frequencies* $(P_1, \ldots, P_9)$ *such that* $(\pi_0, \pi_1, \pi_2)$ *are the ASP IBD probabilities for the strict-recessive model with mating type frequencies* $(P_1, \ldots, P_9)$.

This illustrates the potentially large impact of non-random mating on the IBD probabilities.

*Proof.* The ASP strict-recessive IBD probabilities are independent of $f_2$, so, without loss of generality, we let $f_2 = 1$. These probabilities may be computed using Table 10 and are given by

$$\pi_0 = \frac{P_1}{4P_1 + P_2 + P_3 + P_6/4},$$
$$\pi_1 = \frac{2P_1 + P_2/2 + P_3/2}{4P_1 + P_2 + P_3 + P_6/4},$$
$$\pi_2 = \frac{P_1 + P_2/2 + P_3/2 + P_6/4}{4P_1 + P_2 + P_3 + P_6/4}. \tag{26}$$

Letting

$$P_1 = c\pi_0,$$
$$P_2 = P_3 = c(\pi_1 - 2\pi_0),$$
$$P_6 = 4c(\pi_0 + \pi_2 - \pi_1),$$

where $c$ is a positive constant such that $P_1 + P_2 + P_3 + P_6 \leq 1$, yields $(\pi_0, \pi_1, \pi_2)$. In particular, the boundaries of the ASP triangle are covered by the following models:

- $\pi_0 = 0$: Set $P_1 = 0$ and vary $P_6$ from 0 to 1;
- $\pi_1 = \frac{1}{2}$: Set $P_6 = 0$ and vary $P_1$ from 0 to 1;
- $\pi_1 = 2\pi_0$: Set $P_2 = P_3 = 0$ and vary $P_1$ from 0 to 1.

□

Note that for DSPs $\pi_2 = 0$ and hence the strict-recessive model doesn't cover the DSP possible triangle.

**6.3. Some counterexamples.** The following are two counterexamples demonstrating the necessity of the assumptions of our model for the triangle constraints, and examples of models for which $\pi_{01} \neq \pi_{10}$. Either environmental covariance in the sib phenotypes or over-dominance may lead to IBD probabilities that fall outside the ASP possible triangle.

TABLE 10

*Conditional probability of ASP given inheritance vector $x$ and parental genotype $pg$ for a strict-recessive model, under* `Assumption G1`. *For parental genotypes not listed in the table $pr(ASP|x, pg) = 0$.*

$pr(ASP|x, pg)$

| Mating type $mt$ | Parental genotype $pg$ | Inheritance vector $x$ | | | |
|---|---|---|---|---|---|
| | | (1,3,1,3) | (1,3,1,4) | (1,3,2,3) | (1,3,2,4) |
| 1 | $DD \times DD$ | 1 | 1 | 1 | 1 |
| 2 | $DD \times Dd$ | 1 | 0 | 1 | 0 |
| | $DD \times dD$ | 0 | 0 | 0 | 0 |
| 3 | $Dd \times DD$ | 1 | 1 | 0 | 0 |
| | $dD \times DD$ | 0 | 0 | 0 | 0 |
| 6 | $Dd \times Dd$ | 1 | 0 | 0 | 0 |
| | $Dd \times dD$ | 0 | 0 | 0 | 0 |
| | $dD \times Dd$ | 0 | 0 | 0 | 0 |
| | $dD \times dD$ | 0 | 0 | 0 | 0 |

**6.3.1.** `Assumption M1` - **Monotonicity of penetrances.** Under random mating and Hardy-Weinberg equilibrium, the triangle constraints hold regardless of the penetrances. However, `Assumption M1` is necessary for the triangle constraints when we allow arbitrary mating type frequencies. An extreme example is that of a diallelic model with over-dominance, $f_0 = f_2 = 0, f_1 = 1$, and mating type frequency $P_6 = 1$. Then, for ASPs, $(\pi_0, \pi_1, \pi_2) = (\frac{1}{2}, 0, \frac{1}{2})$.

**6.3.2.** `Assumption G1` - **Conditional independence of phenotypes given genotypes.** Consider a genetic model with random mating and Hardy-Weinberg equilibrium at the DS locus, and the following extreme form of environmental covariance:

$$pr(\text{Both sibs are affected} \mid \text{At least one sib has genotype } DD) = 1,$$

$$pr(\text{Both sibs are affected} \mid \text{Neither of the sibs has genotype } DD) = 0.$$

For this model, the ASP IBD probabilities for $p \neq 0$ are given by

$$\pi_0 = \frac{p^2 + 4p(1-p) + 2(1-p)^2}{4p^2 + 12p(1-p) + 7(1-p)^2},$$

$$\pi_1 = \frac{2p^2 + 6p(1-p) + 4(1-p)^2}{4p^2 + 12p(1-p) + 7(1-p)^2},$$

$$\pi_2 = \frac{p^2 + 2p(1-p) + (1-p)^2}{4p^2 + 12p(1-p) + 7(1-p)^2}.$$

Hence, $2\pi_0 > \pi_1$ and $\pi_1 > \frac{1}{2}$ for $p \neq 1$, and **Assumption G1** is necessary for the triangle constraints.

*Proof.* The result follows from Tables 11 and 12, and equation (1).

TABLE 11

***Conditional probability of ASP given inheritance vector $x$ and parental genotype $pg$ for a model which violates Assumption G1. For parental genotypes not listed in the table $pr(ASP|x, pg) = 0$.***

$pr(ASP|x, pg)$

| Parental genotype | Inheritance vector $x$ | | | |
|---|---|---|---|---|
| $pg$ | (1,3,1,3) | (1,3,1,4) | (1,3,2,3) | (1,3,2,4) |
| $DD \times DD$ | 1 | 1 | 1 | 1 |
| $DD \times Dd$ | 1 | 1 | 1 | 1 |
| $DD \times dD$ | 0 | 1 | 0 | 1 |
| $Dd \times DD$ | 1 | 1 | 1 | 1 |
| $dD \times DD$ | 0 | 0 | 1 | 1 |
| $Dd \times Dd$ | 1 | 1 | 1 | 1 |
| $Dd \times dD$ | 0 | 1 | 0 | 0 |
| $dD \times Dd$ | 0 | 0 | 1 | 0 |
| $dD \times dD$ | 0 | 0 | 0 | 1 |

TABLE 12

| $x$ | $\sum_{pg} pr(ASP\|x, pg) pr(pg)$ |
|---|---|
| (1,3,1,3) | $p^4 + 2p^3q + p^2q^2$ |
| (1,3,1,4) | $p^4 + 3p^3q + 2p^2q^2$ |
| (1,3,2,3) | $p^4 + 3p^3q + 2p^2q^2$ |
| (1,3,2,4) | $p^4 + 4p^3q + 2p^2q^2$ |

□

**6.3.3. Examples when $\pi_{01} \neq \pi_{10}$ .** Either parental imprinting or non-symmetry of parental mating type frequencies with respect to maternal and paternal genotypes may result in $\pi_{01}$ being different from $\pi_{10}$.

First, consider a single diallelic DS locus with random mating and Hardy-Weinberg equilibrium, and assume that $f_{00} = f_{01} = 0$ and $f_{10} = f_{11} = 1$. Then, for ASPs,

$$\sum_{pg} pr(ASP|(1,3,1,4), pg)\, pr(pg) = p^4 + 3p^3q + 3p^2q^2 + pq^3,$$

$$\sum_{pg} pr(ASP|(1,3,2,3), pg)\, pr(pg) = p^4 + 2p^3q + p^2q^2,$$

and hence, by equation (1), $\pi_{01} \neq \pi_{10}$.

Consider now a strict-recessive model with general mating type frequencies and $P_2 \neq P_3$. From the derivation of equation (26) for ASPs

$$\pi_{10} = \frac{P_1 + P_3/2}{4P_1 + P_2 + P_3 + P_6/4} \neq \frac{P_1 + P_2/2}{4P_1 + P_2 + P_3 + P_6/4} = \pi_{01}.$$

Hence, in some cases, distinguishing between maternal and paternal sharing may lead to more powerful tests of linkage.

**6.4. Non-random mating and Hardy-Weinberg disequilibrium models.** In order to investigate the impact of non-random mating and Hardy-Weinberg disequilibrium on the IBD probabilities, we consider the model used by Jin *et al.* [15] for the mating type frequencies at a single $m$-allele DS locus. Let $p_{ij,kl}$ denote the probability of a $D_iD_j \times D_kD_l$ mating, $p_{ij}$ the probability of genotype $D_iD_j$, and $p_i$ the allele frequency of $D_i$. The mating type frequencies are mixtures of the frequencies under random mating and complete dependence, with the same margins. They are given by

$$p_{ij,kl} = \begin{cases} (1-\delta_R)p_{ij}^2 + \delta_R p_{ij}, & \text{if } i = k \text{ and } j = l, \\ (1-\delta_R)p_{ij}p_{kl}, & \text{otherwise,} \end{cases}$$

where

$$p_{ij} = \begin{cases} p_i^2 + \delta_{HW} p_i(1-p_i), & \text{if } i = j, \\ 2p_ip_j(1-\delta_{HW}), & \text{if } i \neq j, \end{cases}$$

and $i \leq j$, $k \leq l$. Here, $\delta_{HW}$ is a parameter representing deviation from Hardy-Weinberg equilibrium, while $\delta_R$ represents deviation from random mating. Positive $\delta_{HW}$ correspond to a deficiency of heterozygotes compared to Hardy-Weinberg frequencies, and similarly, positive $\delta_R$ correspond to a deficiency of different genotype matings compared to random mating frequencies. Louis *et al.* [22] used this model with $\delta_{HW} = 0$ to study the impact of positive assortative mating on ASP IBD probabilities and Weir [37] used this model with $\delta_R = 0$ as a one-parameter class of alternatives to Hardy-Weinberg equilibrium. For the usual two-allele model, with alleles $D$ and $d$ and allele frequencies $p$ and $q = 1-p$, respectively, the mating type frequencies are given in Table 13. The parameters $(\delta_R, \delta_{HW})$ are constrained to yield non-negative genotype and mating type frequencies, in particular, $\delta_R \leq 1$ and $\delta_{HW} \leq 1$.

PROPOSITION 8. Impact of non-random mating on ASP strict-recessive curve.
*For a strict-recessive model ($f_0 = f_1 = 0, f_2 = 1$), the ASP IBD probabilities satisfy the following:*

- *If $\delta_R = 1$, then $\pi_1 = 2\pi_0$.*

TABLE 13
*Mating type frequencies for non-random mating model.*

| Mating type $mt$ | Mating type frequency $P_{mt}$ |
|---|---|
| 1 | $(1-\delta_R)(p^2+\delta_{HW}pq)^2+\delta_R(p^2+\delta_{HW}pq)$ |
| 2 | $(1-\delta_R)(p^2+\delta_{HW}pq)(2pq(1-\delta_{HW}))$ |
| 3 | $(1-\delta_R)(p^2+\delta_{HW}pq)(2pq(1-\delta_{HW}))$ |
| 4 | $(1-\delta_R)(p^2+\delta_{HW}pq)(q^2+\delta_{HW}pq)$ |
| 5 | $(1-\delta_R)(p^2+\delta_{HW}pq)(q^2+\delta_{HW}pq)$ |
| 6 | $(1-\delta_R)(2pq(1-\delta_{HW}))^2+\delta_R(2pq(1-\delta_{HW}))$ |
| 7 | $(1-\delta_R)(2pq(1-\delta_{HW}))(q^2+\delta_{HW}pq)$ |
| 8 | $(1-\delta_R)(2pq(1-\delta_{HW}))(q^2+\delta_{HW}pq)$ |
| 9 | $(1-\delta_R)(q^2+\delta_{HW}pq)^2+\delta_R(q^2+\delta_{HW}pq)$ |

- *When $\delta_R = 0$, i.e. mating is random, then regardless of $\delta_{HW}$, the ASP strict-recessive IBD probabilities lie on the Hardy-Weinberg curve $\pi_1^2 = 4\pi_0\pi_2$.*
- *When $0 < \delta_R \leq 1$, the ASP strict-recessive IBD probabilities are between the Hardy-Weinberg curve and the line $\pi_1 = 2\pi_0$, i.e. $\pi_1^2 \leq 4\pi_0\pi_2$ and $\pi_1 \geq 2\pi_0$. Also, if $0 \leq \delta_{HW} \leq 1$*

$$\lim_{p\to 1}(\pi_0,\pi_1,\pi_2) = \left(\frac{1}{4},\frac{1}{2},\frac{1}{4}\right),$$

$$\lim_{p\to 0, p\neq 0}(\pi_0,\pi_1,\pi_2) = \frac{1}{\frac{7}{2}\delta_{HW}+\frac{1}{2}}\left(\delta_{HW}, 2\delta_{HW}, \frac{1}{2}(\delta_{HW}+1)\right),$$

*hence the limit of the IBD probabilities as $p \to 0$ is on the line $\pi_1 = 2\pi_0$ and is independent of $\delta_R$.*
- *When $\delta_R < 0$, then regardless of $\delta_{HW}$, the ASP strict-recessive IBD probabilities are between the Hardy-Weinberg curve and the line $\pi_1 = \frac{1}{2}$, i.e. $\pi_1^2 \geq 4\pi_0\pi_2$ and $\pi_1 \leq \frac{1}{2}$.*

These observations are illustrated in Figures 6 and 7 p. 215.

*Proof.* For the strict-recessive model, the ASP probabilities are given by equation (26). Since $P_3 = P_2$, then

$$4\pi_0\pi_2 - \pi_1^2 = \frac{P_1P_6 - P_2^2}{(4P_1+2P_2+P_6/4)^2} = \frac{\delta_R p_{DD} p_{Dd}(1-p_{dd}+p_{dd}\delta_R)}{(4P_1+2P_2+P_6/4)^2},$$

and we have the following cases:
- $\delta_R = 0$: $\pi_1^2 = 4\pi_0\pi_2$.
- $0 < \delta_R \leq 1$: $\pi_1^2 \leq 4\pi_0\pi_2$.

- $\delta_R < 0$: Since $P_1 \geq 0$, then $\delta_R \geq \frac{-p_{DD}}{1-p_{DD}}$ and hence $1 - p_{dd} + p_{dd}\delta_R \geq \frac{1-p_{DD}-p_{dd}}{1-p_{DD}} \geq 0$. It follows that $\pi_1^2 \geq 4\pi_0\pi_2$.
- $\delta_R = 1$: $P_2 = P_3 = 0$ and it follows that $\pi_1 = 2\pi_0$. □

**7. Multilocus conditional distribution of inheritance vectors at marker loci linked to the DS loci given phenotype vector of sibship.** Consider $L$ unlinked markers, $\mathcal{M}^1, \ldots, \mathcal{M}^L$, where $\mathcal{M}^l$ is linked to $\mathcal{D}^l$ at recombination fraction $\theta_l$, $l = 1, \ldots, L$. Let $y^l$ denote the inheritance vector of the sibship at $\mathcal{M}^l$, $l = 1, \ldots, L$, and let $y = (y^1, \ldots, y^L)$. We wish to compute $pr(y|\phi)$, the conditional distribution of the inheritance vectors at the marker loci given the phenotype vector of the sibship. This distribution is obtained by conditioning on all possible recombination patterns in the sibship between the markers and the DS loci. Since the inheritance vectors at the marker loci are conditionally independent of the phenotype vector given the inheritance vectors at the DS loci, then

$$pr(y|\phi) = \sum_x pr(y, x|\phi) = \sum_x pr(y|x, \phi)\; pr(x|\phi) = \sum_x pr(y|x)\; pr(x|\phi).$$

Now, the number of coordinates at which $x^l$ and $y^l$ differ, $\Delta(x^l, y^l) = \sum_{i=1}^{2k}(1 - \delta(x_i^l, y_i^l))$, is the total number of recombinants between $\mathcal{D}^l$ and $\mathcal{M}^l$. The chance that a coordinate differs between $x^l$ and $y^l$ is the chance of a recombination between $\mathcal{D}^l$ and $\mathcal{M}^l$, i.e. the recombination fraction $\theta_l$. Since recombination events are independent for unlinked loci and across meioses, then

$$\begin{aligned} pr(y|\phi) &= \sum_x \left\{ \prod_{l=1}^{L} pr(y^l|x^l) \right\} pr(x|\phi) \\ &= \sum_x \left\{ \prod_{l=1}^{L} \theta_l^{\Delta(x^l,y^l)} (1-\theta_l)^{2k-\Delta(x^l,y^l)} \right\} pr(x|\phi). \end{aligned}$$

Hence, the conditional distribution $pr(y|\phi)$ of inheritance vectors at markers linked to DS loci in the manner described above may be obtained from the conditional distribution $pr(x|\phi)$ of the inheritance vectors at the DS loci by means of the transition matrix

$$T(\theta_1, \ldots, \theta_L) = T(\theta_1) \otimes \ldots \otimes T(\theta_L).$$

$T(\theta_l)$ is the Kronecker power of the $2 \times 2$ transition matrices corresponding to transitions in each of the $2k$ coordinates between $\mathcal{D}^l$ and $\mathcal{M}^l$

$$T(\theta_l) = \begin{bmatrix} 1-\theta_l & \theta_l \\ \theta_l & 1-\theta_l \end{bmatrix}^{\otimes 2k}.$$

This matrix representation separates the contributions of the genetic model for disease susceptibility ($pr(x|\phi)$) and of linkage ($\theta$'s). Note that

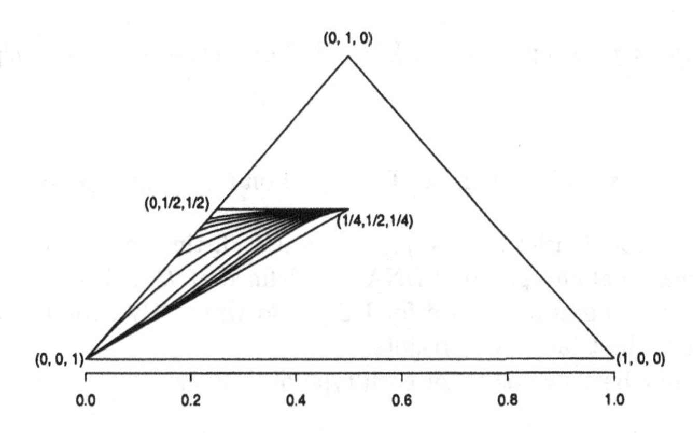


FIG. 6. *ASP strict-recessive curves for non-random mating model with* $\delta_{HW} = 0$ *and* $\delta_R = -0.5, -0.4, \ldots, 0.5$ *(from top to bottom). For fixed* $\delta_R$*, as* $p \to 1$*, the ASP IBD probabilities move along a curve toward* $(\frac{1}{4}, \frac{1}{2}, \frac{1}{4})$.

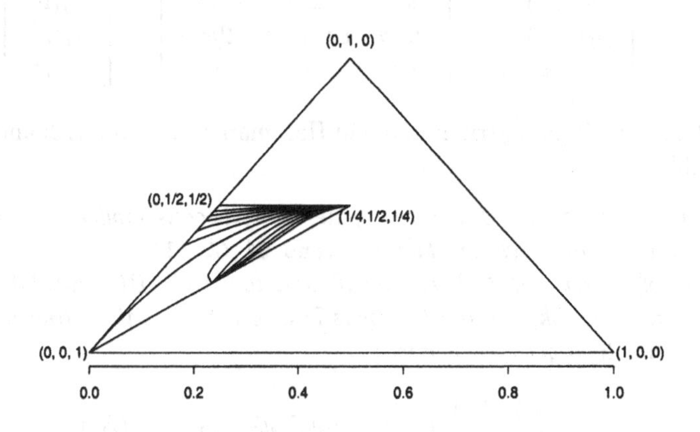


FIG. 7. *ASP strict-recessive curves for non-random mating model with* $\delta_{HW} = 0.1$ *and* $\delta_R = -0.5, -0.4, \ldots, 0.5$ *(from top to bottom). For fixed* $\delta_R$*, as* $p \to 1$*, the ASP IBD probabilities move along a curve toward* $(\frac{1}{4}, \frac{1}{2}, \frac{1}{4})$.

$T(0)$ is the $2^{2k} \times 2^{2k}$ identity matrix and the entries of $T(\frac{1}{2})$ are all equal to $\frac{1}{2^{2k}}$.

Let us now consider the simple case of sib-pairs and a single marker $\mathcal{M}$ linked to a DS locus $\mathcal{D}$ at recombination fraction $\theta$ ($\mathcal{D}$ could be one of several unlinked DS loci). For ASPs and $i = 0, 1, 2$, let

$$\pi_i^{ASP}(\nu) = pr(\text{Sib-pair shares DNA IBD on } i \text{ chromosomes at } \mathcal{D}|ASP)$$

and let

$$\rho_i^{ASP}(\theta, \nu) = pr(\text{Sib-pair shares DNA IBD on } i \text{ chromosomes at } \mathcal{M}|ASP).$$

The IBD probabilities $\pi_{ij}$ and $\rho_{ij}$, $i, j = 0, 1$, distinguishing between sharing of maternal and paternal DNA, are defined at $\mathcal{D}$ and $\mathcal{M}$ as in Section 4. The same notation is used for DSPs and USPs. By Proposition 5, the $\pi$'s satisfy the triangle constraints.

It may be shown that for each type of sib-pair

$$\begin{bmatrix} \rho_{00}(\theta,\nu) \\ \rho_{01}(\theta,\nu) \\ \rho_{10}(\theta,\nu) \\ \rho_{11}(\theta,\nu) \end{bmatrix} = \begin{bmatrix} \psi^2 & \psi\bar{\psi} & \psi\bar{\psi} & \bar{\psi}^2 \\ \psi\bar{\psi} & \psi^2 & \bar{\psi}^2 & \psi\bar{\psi} \\ \psi\bar{\psi} & \bar{\psi}^2 & \psi^2 & \psi\bar{\psi} \\ \bar{\psi}^2 & \psi\bar{\psi} & \psi\bar{\psi} & \psi^2 \end{bmatrix} \times \begin{bmatrix} \pi_{00}(\nu) \\ \pi_{01}(\nu) \\ \pi_{10}(\nu) \\ \pi_{11}(\nu) \end{bmatrix}, \tag{27}$$

where $\psi = \theta^2 + (1-\theta)^2$ and $\bar{\psi} = 1 - \psi = 2\theta\bar{\theta}$. When we do not distinguish between maternal and paternal sharing, the transition matrix $T(\theta)$ collapses into a $3 \times 3$ matrix

$$\begin{bmatrix} \rho_0(\theta,\nu) \\ \rho_1(\theta,\nu) \\ \rho_2(\theta,\nu) \end{bmatrix} = \begin{bmatrix} \psi^2 & \psi\bar{\psi} & \bar{\psi}^2 \\ 2\psi\bar{\psi} & \psi^2 + \bar{\psi}^2 & 2\psi\bar{\psi} \\ \bar{\psi}^2 & \psi\bar{\psi} & \psi^2 \end{bmatrix} \times \begin{bmatrix} \pi_0(\nu) \\ \pi_1(\nu) \\ \pi_2(\nu) \end{bmatrix}. \tag{28}$$

This $3 \times 3$ transition matrix is given in Haseman and Elston [12] and Suarez *et al.* [35].

PROPOSITION 9. *Under our general multilocus model with arbitrary mating type frequencies and* `Assumptions G1,G2, M2`

- $(\rho_0^{ASP}(\theta,\nu), \rho_1^{ASP}(\theta,\nu), \rho_2^{ASP}(\theta,\nu))$, *the ASP IBD probabilities at a marker linked to the DS locus $\mathcal{D}$ at a recombination fraction $\theta$, fall in a triangle with vertices*

$$\left(\frac{1}{4}, \frac{1}{2}, \frac{1}{4}\right), \quad (\bar{\psi}^2, 2\psi\bar{\psi}, \psi^2) \text{ and } \frac{1}{2}(\bar{\psi}, 1, \psi)$$

*where* $\psi = \theta^2 + (1-\theta)^2$;
- *this triangle is contained in the ASP possible triangle* ($\theta = 0$);
- *as* $\theta \to \frac{1}{2}$ *the triangles shrink toward* $(\frac{1}{4}, \frac{1}{2}, \frac{1}{4})$ *along the line* $\rho_1 = \frac{1}{2}$.

*Similarly,*

- $(\rho_0^{DSP}(\theta,\nu), \rho_1^{DSP}(\theta,\nu), \rho_2^{DSP}(\theta,\nu))$, *the DSP IBD probabilities at a marker linked to the DS locus* $\mathcal{D}$ *at a recombination fraction* $\theta$, *fall in a triangle with vertices*

$$\left(\frac{1}{4},\frac{1}{2},\frac{1}{4}\right),\quad \frac{1}{2}(\psi,1,\bar{\psi}) \text{ and } \frac{1}{3}(1-\bar{\psi}^2, 2(\psi\bar{\psi}+\psi^2+\bar{\psi}^2), 1-\psi^2);$$

- *this triangle is contained in the DSP possible triangle* $(\theta = 0)$;
- *as* $\theta \to \frac{1}{2}$ *the triangles shrink toward* $(\frac{1}{4},\frac{1}{2},\frac{1}{4})$ *along the line* $\rho_1 = \frac{1}{2}$.

*Again, USP IBD probabilities satisfy the same constraints as ASP IBD probabilities.*

*Proof.* For ASPs (USPs), the proof relies on $\frac{1}{2} \leq \psi \leq 1$, on equations (7) and (8), and on the relationship between IBD probabilities at the marker and at the DS locus

$$\begin{bmatrix} \rho_0 \\ \rho_1 \\ \rho_2 \end{bmatrix} = \pi_0 \begin{bmatrix} \psi^2 \\ 2\psi\bar{\psi} \\ \bar{\psi}^2 \end{bmatrix} + \pi_1 \begin{bmatrix} \psi\bar{\psi} \\ \psi^2+\bar{\psi}^2 \\ \psi\bar{\psi} \end{bmatrix} + \pi_2 \begin{bmatrix} \bar{\psi}^2 \\ 2\psi\bar{\psi} \\ \psi^2 \end{bmatrix}, \tag{29}$$

whereby $(\rho_0,\rho_1,\rho_2)$ is a convex combination of 3 trinomial probability vectors contained in the triangle with vertices (1,0,0), (0,1,0), (0,0,1). The proof is similar for DSPs. □

The possible triangles are shown in Figure 8 for various values of the recombination fraction $\theta$. Figure 9 shows the impact of recombination on the IBD probabilities for four models. The following can easily be shown with representation (29):

- If $\pi_1^2 = 4\pi_0\pi_2$, then $\rho_1^2 = 4\rho_0\rho_2$. Hence, strict-recessive random mating ASP IBD probabilities at the marker $\mathcal{M}$ remain on the Hardy-Weinberg curve.
- If $\pi_1 = \frac{1}{2}$, then $\rho_1 = \frac{1}{2}$. Hence additive probabilities at the marker $\mathcal{M}$ remain on the additive line.

**8. Open questions.** In this paper, we studied sib-pair IBD probabilities under general multilocus models for disease susceptibility. We proved the possible triangle constraints under general monotonicity assumptions concerning the penetrances, and without the problematic assumptions of random mating, Hardy-Weinberg equilibrium and linkage equilibrium. We also studied the parameterization of sib-pair IBD probabilities for common genetic models and showed that the ASP possible triangle was covered by the IBD probabilities of at least two types of models. Finally, we illustrated the potentially large impact of non-random mating on IBD probabilities.

Several open questions regarding IBD probabilities come to mind. Firstly, the problem of *linked* DS loci remains to be addressed. An obvious complication in the derivation of constraints for this type of models arises since $pr(x|pg)$ now involves recombination fractions between the linked DS loci and hence does not cancel out of equation (1).

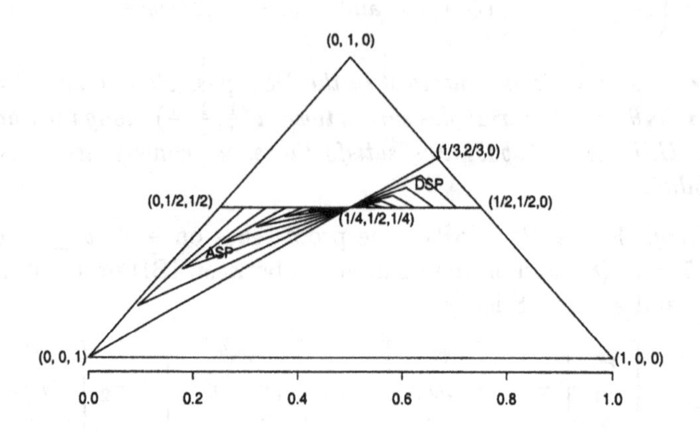


FIG. 8. *ASP and DSP possible triangles for IBD probabilities at a marker $\theta$ away from a DS locus, $\theta = 0, 0.05, \ldots, 0.5$.*

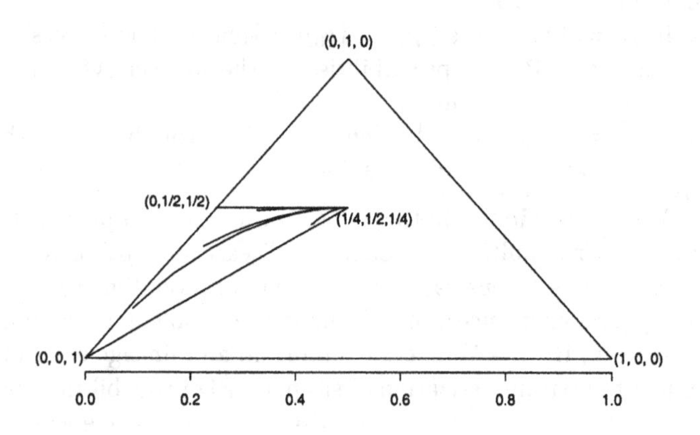


FIG. 9. *Curves traced by ASP IBD probabilities at a marker, as recombination fraction $\theta$ between the marker and a DS locus varies between 0 and $\frac{1}{2}$. The 4 models considered involve a single diallelic random mating DS locus. Starting from the top curve, the models are: strict-dominant with $p = 0.1$, intermediate with $s = 0.1$, $p = 0.1$, strict-recessive with $p = 0.1$, and quasi-recessive with $r = 10$, $p = 0.1$.*

This invalidates Proposition 1 for linked loci. A second open problem is the coverage of the space for *multilocus* IBD probabilities. For two loci, for example, what region of (3×3)-1–dimensional space is covered by $\pi_{i_1,i_2}$, the probability that an ASP shares DNA IBD on $i_1$ and $i_2$ chromosomes at locus 1 and 2, respectively? A third question is the generalization of the constraints to larger sibships with both affected and unaffected individuals. We have already derived constraints for affected sib-trio IBD probabilities, however generalizing the constraints to arbitrary sibships is challenging.

We used the matrix representation of the IBD probabilities at markers linked to DS loci (transition matrix $T(\theta)$) to derive a score test of the null hypothesis of no linkage between a marker and a DS locus for sibships of various sizes and phenotype patterns [7]. This test is locally most powerful in the recombination fraction $\theta$ between the marker and a DS locus, and provides optimal weights for combining the test statistics across the different types of sibships. For affected only sibships, the score statistic is the usual statistic $S_{pairs}$, obtained by forming all possible pairs of affected sibs and averaging the proportions of chromosomes on which they share DNA IBD. The matrix representation and score tests may be extended to other types of pedigrees.

**Acknowledgments.** We would like to thank David Nelson for kindly providing the program for barycentric plots as well as the simplex figure. We are also grateful to participants at the IMA meeting at which a version of these results were presented. Their lively discussion spurred us to further develop some of our ideas on age and sex-dependent penetrances. Finally, we wish to thank the referee for helpful comments. This work was supported in part by a PMMB Burroughs-Wellcome Fellowship, and by the NIH through grant RO1 HG01093.

## REFERENCES

[1] SCHIZOPHRENIA COLLABORATIVE LINKAGE GROUP (CHROMOSOME 22), A combined analysis of D22S278 marker alleles in affected sib-pairs: support for a susceptibility locus for schizophrenia at chromosome 22q12. *Am. J. Med. Genet.*, 67:40–45, 1996.

[2] W.C. BLACKWELDER AND R.C. ELSTON, A comparison of sib-pair linkage tests for disease susceptibility loci. *Genet. Epidemiol.*, 2:85–97, 1985.

[3] H.J. CORDELL, J.A. TODD, S.T. BENNETT, Y. KAWAGUCHI, AND M. FARRALL, Two-locus maximum lod score analysis of a multifactorial trait: Joint consideration of IDDM2 and IDDM4 with IDDM1 in type I diabetes. *Am. J. Hum. Genet.*, 57:920–934, 1995.

[4] N.J. COX AND R.S. SPIELMAN, The insulin gene and susceptibility to IDDM. *Genet. Epidemiol.*, 6:65–69, 1989.

[5] A.G. CUDWORTH AND J.C. WOODROW, Evidence for HLA-linked genes in "juvenile" diabetes mellitus. *Br. Med. J.*, 3:133–135, 1975.

[6] N.E. DAY AND M.J. SIMONS, Disease-susceptibility genes — their identification by multiple case family studies. *Tissue Antigens*, 8:109–119, 1976.

[7] S. DUDOIT AND T.P. SPEED, A score test for linkage using identity by descent data from sibships. In preparation.

[8] S.N. ETHIER AND S.E. HODGE, Identity-by-descent analysis of sibship configurations. *Am. J. Med. Genet.*, 22:263–272, 1985.

[9] J.J. FARAWAY, Improved sib-pair linkage test for disease susceptibility loci. *Genet. Epidemiol.*, 10:225–233, 1993.

[10] E. FEINGOLD, P.O. BROWN, AND D. SIEGMUND, Gaussian models for genetic linkage analysis using complete high-resolution maps of identity by descent. *Am. J. Hum. Genet.*, 53:234–251, 1993.

[11] E. FEINGOLD AND D. SIEGMUND, Strategies for mapping heterogeneous recessive traits by allele sharing methods. *Am. J. Hum. Genet.*, 60:965–978, 1997.

[12] J.K. HASEMAN AND R.C. ELSTON, The investigation of linkage between a quantitative trait and a marker locus. *Behavior Genetics*, 2:3–19, 1972.

[13] P. HOLMANS, Asymptotic properties of affected-sib-pair linkage analysis. *Am. J. Hum. Genet.*, 52:362–374, 1993.

[14] P. HOLMANS AND D. CLAYTON, Efficiency of typing unaffected relatives in an affected-sib-pair linkage study with single-locus and multiple tightly linked markers. *Am. J. Hum. Genet.*, 57:1221–1232, 1995.

[15] K. JIN, T.P. SPEED, AND G. THOMSON, Tests of random mating for a highly polymorphic locus: Application to HLA data. *Biometrics*, 51:1064–1076, 1995.

[16] M. KNAPP, S.A. SEUCHTER, AND M.P. BAUR, Linkage analysis in nuclear families 1: Optimality criteria for affected sib-pair tests. *Hum. Hered.*, 44:37–43, 1994.

[17] M. KNAPP, S.A. SEUCHTER, AND M.P. BAUR, Linkage analysis in nuclear families 2: Relationship between affected sib-pair tests and lod score analysis. *Hum. Hered.*, 44:44–51, 1994.

[18] M. KNAPP, S.A. SEUCHTER, AND M.P. BAUR, Two-locus disease models with two marker loci: The power of affected sib-pair tests. *Am. J. Hum. Genet.*, 55:1030–1041, 1994.

[19] L. KRUGLYAK, M.J. DALY, M.P. REEVE-DALY, AND E.S. LANDER, Parametric and nonparametric linkage analysis: a unified multipoint approach. *Am. J. Hum. Genet.*, 58:1347–1363, 1996.

[20] L. KRUGLYAK AND E.S. LANDER, Complete multipoint sib-pair analysis of qualitative and quantitative traits. *Am. J. Hum. Genet.*, 57:439–454, 1995.

[21] M. LALANDE, Parental imprinting and human disease. *Annu. Rev. Genet.*, 30:173–195, 1997.

[22] E.J. LOUIS, H. PAYAMI, AND G. THOMSON, The affected sib method. V. Testing the assumptions. *Ann. Hum. Genet.*, 51:75–92, 1987.

[23] E.J. LOUIS, G. THOMSON, AND H. PAYAMI, The affected sib method. II. The intermediate model. *Ann. Hum. Genet.*, 47:225–243, 1983.

[24] M.F. MOFFAT, P.A. SHARP, J.A. FAUX, R.P. YOUNG, W.O.C.M. COOKSON, AND J.M. HOPKIN, Factors confounding genetic linkage between atopy and chromosome 11q. *Clin. Exp. Allergy*, 22:1046–1051, 1992.

[25] G. MORAHAN, D. HUANG, B.D. TAIT, P.G. COLMAN, AND L.C. HARRISON, Markers on distal chromosome 2q linked to insulin-dependent diabetes mellitus. *Science*, 272:1811–1813, 1996.

[26] U. MOTRO AND G. THOMSON, The affected sib method. I. Statistical features of the affected sib-pair method. *Genetics*, 110:525–538, 1985.

[27] N. NIIKAWA, Genomic imprinting and its relevance to genetics diseases. *Jpn. J. Human Genet.*, 41:351–361, 1996.

[28] H. PAYAMI, E.J. LOUIS G. THOMSON, U. MOTRO, AND E. HUDES, The affected sib method. IV. Sib trios. *Ann. Hum. Genet.*, 49:303–314, 1985.

[29] H. PAYAMI, G. THOMSON, AND E.J. LOUIS, The affected sib method. III. Selection and recombination. *Am. J. Hum. Genet.*, 36:352–362, 1984.

[30] M.A. PERICAK-VANCE, J.L. BEBOUT, P.C. GASKELL, JR., L.H. YAMAOKA, W.Y. HUNG, M.J. ALBERTS, A.P. WALKER, R.J. BARTLETT, C.A. MAYNES, K.A. WELSH, N.L. EARL, A. HEYMAN, C.M. CLARK, AND A.D. ROSES, Linkage studies in familial Alzheimer disease: evidence for chromosome 19 linkage. *Am. J. Hum. Genet.*, 48:1034–1050, 1991.

[31] N. Risch, Linkage strategies for genetically complex traits. I. Multilocus models. *Am. J. Hum. Genet.*, 46:222–228, 1990.
[32] N. Risch, Linkage strategies for genetically complex traits. II. The power of affected relative pairs. *Am. J. Hum. Genet.*, 46:229–241, 1990.
[33] N. Risch, Linkage strategies for genetically complex traits. III. The effect of marker polymorphism on analysis of affected relative pairs. *Am. J. Hum. Genet.*, 46:242–253, 1990.
[34] R.S. Spielman, L. Baker, and C.M. Zmijewski, Gene dosage and susceptibility to insulin-dependent diabetes. *Ann. Hum. Genet., Lond.*, 44:135–150, 1980.
[35] B.K. Suarez, J. Rice, and T. Reich, The generalized sib-pair IBD distribution: its use in the detection of linkage. *Ann. Hum. Genet., Lond.*, 42:87–94, 1978.
[36] G. Thomson and W.F. Bodmer, The genetics of HLA and disease associations. In J. Dausset and A. Svejgaard, editors, *HLA and Disease*, pages 545–563. Munksgaard, Copenhagen, 1977.
[37] B.S. Weir, *Genetic Data Analysis II.* Sinauer Associates, 1996.

# A STATISTICAL ANALYSIS OF CANCER GENOME VARIATION

M.A. NEWTON, T.R. YEAGER, AND C.A. REZNIKOFF*

**Abstract.** We consider separating signal and noise in data measuring genetic abnormalities of cancer tumor cells. The linked hot-cold model is introduced and fit to data from a comparative genomic hybridization study of bladder cancer. We analyze these data and report details of the calculation summarized in Yeager *et al.* 1998. The model-generating framework developed in Newton *et al.* 1998 is used to motivate the linked hot-cold model.

**Key words.** Comparative genomic hybridization, deletions, instability-selection model, linked hot-cold model, Poisson process, statistical genetics, suppressor gene.

**1. Introduction.** Recent advances in molecular technology have enabled oncologists to identify a wide range of genetic abnormalities present in cancer cells. The pattern of observed abnormalities, such as the deletion or amplification of particular genomic regions, provides some information about pathways to the cancer phenotype, but inference about such pathways is complicated by substantial intrinsic variation. Cells of clinically indistinguishable tumors can present different genetic abnormalities. For the most part, observed abnormalities are somatic rather than inherited, a fact which separates the analysis of such data from standard statistical genetics. Evidence supports the view that genetic instability, however initiated, creates cells having abnormal genomes, and that selection pressures dictate which cell lineages are likely to survive as tumors. Multiple genetic changes characterize tumorigenesis, and multiple pathways may lead to the same endpoint. Of course, there are numerous and complex biochemical details within this general framework.

Simple statistical methods can assist the oncologist in separating biologically relevant signals from the noise of genetic instability. For example, it may be useful to know if the number of genetic deletions observed in some region is more than would be expected by chance. Data types vary, but in our experience data often exhibit structural properties which render inappropriate or inefficient the use of standard statistical methods. We take the position that statistical methods based on simple probability models are naturally justified. A system for constructing such stochastic models was developed in [9], and we exercise it here to motivate an analysis of data from a study of bladder cancer [12].

Figure 1, reproduced from [12], shows losses and gains of genetic material, relative to normal, in six late stage transitional cell carcinomas of the

*Department of Statistics and Department of Biostatistics and Medical Informatics (MAN); Department of Human Oncology (TRY and CAR), University of Wisconsin-Madison, Madison, WI. MAN was supported by grant R29 CA64364-02 from the National Cancer Institute and CAR was supported by NIH-CA29525-12.

bladder. Together with an array of additional biomolecular information, these data support a particular hypothesis about the pathogenesis of bladder cancer. The two *chromagrams* in Figure 1 themselves are a summary of a comparative genomic hybridization (CGH) study of the six cancers. CGH measures the amount of a particular DNA segment in the tumor cells compared to the normal cells of a given donor, simultaneously for a library of DNA segments [5]. Genomic regions which have been deleted during tumor development are identified by noting a reduction in the ratio of tumor to normal signal, while amplified regions show an increase in this ratio. CGH measures both losses and gains on the scale of chromosome bands, though we reduce the data to events at the chromosome-arm level in our analysis. (By contrast, allelic-loss studies characterize deletions at a finer scale of molecular markers.)

One way to process the CGH data is to form chromagrams like those in Figure 1. It makes biological sense to separate losses and gains. Focusing on one kind of change, still we have at least five distinct outcomes for a given non-acrocentric chromosome. Let us consider losses, and recall that each non-acrocentric chromosome is partitioned into two chromosome arms; the $p-$arm and the $q-$arm, joined at the centromere. Acrocentric chromosomes 13,14,15,21, & 22 are basically just $q-$arms. The five possibilities are: $S_1$ no loss on either arm; $S_2$ the loss of an entire chromosome copy (what we will call a linked loss); $S_3$ loss only on the $p-$arm; $S_4$ loss only on the $q-$arm; and $S_5$ unlinked losses on the $p$ and $q-$arms. Information in the chromagrams of Figure 1 is separated vertically into $p-$arm and $q-$arm data, and shaded boxes indicate losses (gains, resp.). The horizontal hash marks mean that the change is linked to one on the opposite arm; i.e., it is a realization of $S_2$. The identity of tumors is not recorded in the chromagram, so we cannot distinguish state $S_5$ from states $S_3$ or $S_4$, but the raw data enable us to make this distinction. Roughly speaking, the scientific question is whether the changes observed in Figure 1 are consistent with some notion of sporadic change, or if certain arms exhibit elevated rates of change.

The linkage information complicates the statistical analysis, and we could ignore it at the risk of losing statistical efficiency, but we chose not to. Already we are ignoring fine-structure information in CGH data which characterizes changes at the resolution of chromosome bands. Some combination of the methods in [9] and [2] may prove to be effective, though we leave this to future consideration. The present paper focuses on models for CGH data summarizing changes at the level of chromosome arms.

**2. Linked Hot-Cold Model.** We specify a multinomial probability vector for the five basic states of a non-acrocentric chromosome and the two states of an acrocentric chromosome using a small set of interpretable parameters. These parameters control departures from the null hypothesis of sporadic change, and thus enable a test of this null hypothesis against the alternative that certain regions exhibit elevated rates of change. Fur-

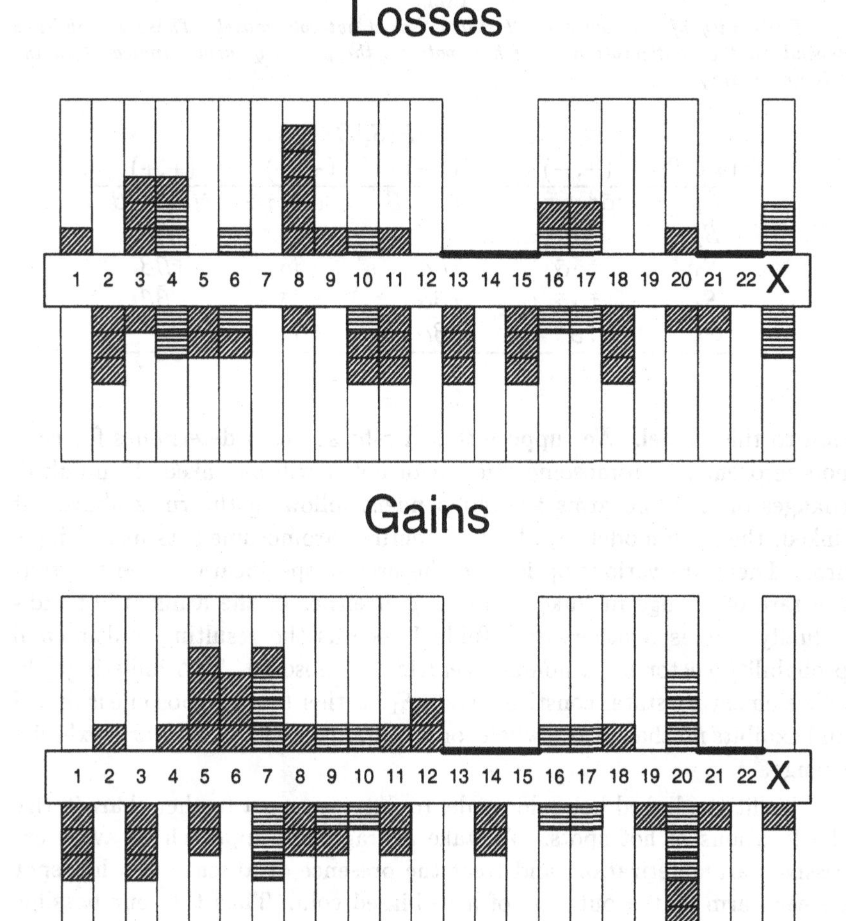


FIG. 1. *Chromagrams showing genetic abnormalities as measured by CGH in six bladder cancers: Shaded boxes count the number of tumors (out of the sample of 6) with losses (top panel) or gains (lower panel) over the 23 chromosome pairs. The vertical axis is separated into p−arm abnormalities (above the numbers) and q−arm abnormalities (below the numbers). Horizontal hashing indicates that the change is linked to one on the opposite arm (see text). Unshaded volume is proportional to the number of tumors exhibiting no CGH change on that chromosome arm.*

thermore, conditional probability calculations within this model lead to an extraction of signal from multinomial data in Figure 1.

The linked hot-cold model says that each chromosome arm is either hot or cold, which means, roughly, that hot arms exhibit change at a rate $\beta$ which is elevated compared to the background rate of change, say $\alpha$, the characteristic rate for cold arms. Linkage creates a slight complica-

TABLE 1

*Probability of observable CGH states, linked hot-cold model. These probabilities depend on the configuration G of hot-spots on the p and q−arms, indicated in the column heading.*

| | $P(S_i \mid G)$ | | | |
|---|---|---|---|---|
| State $S_i$ | $(-,-)$ | $(+,-)$ | $(-,+)$ | $(+,+)$ |
| $S_1$ | $\bar{\tau}\bar{\alpha}^2 + \tau\bar{\alpha}$ | $\bar{\tau}\beta\bar{\alpha} + \tau\beta$ | $\bar{\tau}\beta\bar{\alpha} + \tau\bar{\alpha}$ | $\bar{\tau}\beta^2 + \tau\beta$ |
| $S_2$ | $\tau\alpha$ | $\tau\beta$ | $\tau\beta$ | $\tau\beta$ |
| $S_3$ | $\bar{\tau}\alpha\bar{\alpha}$ | $\bar{\tau}\beta\bar{\alpha}$ | $\bar{\tau}\bar{\beta}\alpha$ | $\bar{\tau}\bar{\beta}\beta$ |
| $S_4$ | $\bar{\tau}\alpha\bar{\alpha}$ | $\bar{\tau}\bar{\beta}\alpha$ | $\tau\bar{\beta}\bar{\alpha}$ | $\bar{\tau}\bar{\beta}\beta$ |
| $S_5$ | $\bar{\tau}\alpha^2$ | $\bar{\tau}\beta\alpha$ | $\bar{\tau}\beta\alpha$ | $\bar{\tau}\beta^2$ |

tion to this model. We suppose that a $\tau$-biased coin determines for each non-acrocentric chromosome whether or not it will be linked. If unlinked, changes on the two arms are independent following the rules above. If linked, then the model says that the entire chromosome acts like a single arm. There are various options in the precise specification. We say that the rate of change in linked arms is $\beta$ if either of the arms is hot individually, and is $\alpha$ otherwise. Table 1 records the resulting multinomial probability vector for a non-acrocentric chromosome. Two indistinguishable elementary states constitute state $S_1$. Either the chromosome is linked and exhibits no change as a whole, or it is unlinked and neither arm exhibits change.

To fit the linked hot-cold model to data, we must further characterize the locations of hot spots. We take a pragmatic approach to avoid excessive parameterization, and treat the presence or absence of a hot spot on each arm as the outcome of a $\eta$−biased coin. Thus the four possible configurations $G = (G_p, G_q)$ of hot spots on $p$ and $q$ have probabilities $P(G)$: $\bar{\eta}^2, \bar{\eta}\eta, \bar{\eta}\eta$, and $\eta^2$ respectively. On acrocentric chromosomes, only two configurations are possible: hot or cold. By treating $G_p$ and $G_q$ as random variables, it is as if the hot spot configurations are missing data.

Maximum likelihood estimates of $\alpha$, $\beta$, $\tau$, and $\eta$ are readily obtained in the hierarchical model formulation. The likelihood function $L$ is the probability of observed data as a function of unknown parameters. Assuming independent changes among chromosomes, independent tumors, and noting the mixing created by unknown hot-spot locations, $L$ becomes

$$L = \prod_{\text{chrom}} \left\{ \sum_{G} P(\text{data}_{\text{chrom}} \mid G)\, P(G) \right\}.$$

The product is over the 23 chromosome pairs, and for each we sum over the four possible configurations $G = (G_p, G_q)$ of hot spots. Acrocentric chromosomes have two terms instead of four. The conditional probability

TABLE 2
*Likelihood Analysis of Losses.*

| hypothesis | parameter estimates $\hat{\alpha}$ | $\hat{\beta}$ | $\hat{\tau}$ | $\hat{\eta}$ | loglikelihood | p-value |
|---|---|---|---|---|---|---|
| sporadic | .22 | – | .28 | – | -125.3 | 1/250 |
| linked hot-cold | .05 | .38 | .22 | .47 | -120.5 | |

TABLE 3
*Likelihood Analysis of Gains.*

| hypothesis | parameter estimates $\hat{\alpha}$ | $\hat{\beta}$ | $\hat{\tau}$ | $\hat{\eta}$ | loglikelihood | p-value |
|---|---|---|---|---|---|---|
| sporadic | .22 | – | .34 | – | -128.9 | 1/250 |
| linked hot-cold | .13 | .48 | .30 | .22 | -125.6 | |

$P(\text{data}_{\text{chrom}}|G)$ is a multinomial probability

$$P(\text{data}_{\text{chrom}}|G) = \prod_{i=1}^{5} p_i^{n_i}$$

with factors $p_i$ as in one of the columns from Table 1. The multinomial count $n_i$ gives the number of instances of state $S_i$ in the set of tumors. For acrocentric chromosomes, the multinomial probability reduces to a product over the two observable states instead of a product over five states.

Of course, for numerical reasons we work with the logarithm of $L$. Optimization over the parameter set is done twice, first under the null hypothesis of sporadic change, $\alpha = \beta$, and then under the alternative hypothesis $\alpha < \beta$. It is interesting that the null hypothesis has two parameters $\alpha$ and $\tau$ while the alternative has four. That is, $\alpha = \beta$ means that hot spots are just like cold spots, and so the probability of data is the same for all configurations of hot spots in this case, providing no information about $\eta$. Since the hot-spot configurations $G$ are missing data, one might expect that the EM algorithm would be helpful, but it is quite simple to use built-in optimization code `nlminb` within the *S-PLUS* computing environment [10].

Tables 2,3 show the numerical results (p-values only were reported in [12]). For example, we estimate that about 47% of the chromosome arms exhibit an elevated loss rate of 38% compared to the background rate of 5%. Linkage occurs about 20% of the time, and this estimate is somewhat insensitive to the hypothesis. Of course, the estimate $\hat{\alpha}$ is higher under the sporadic change hypothesis than under the alternative hypothesis so as to account for chromosomes showing high observed rates of change.

Difference in maximized loglikelihood forms a natural test statistic, but it is unclear how it should be calibrated. Ordinarily, we would appeal to Wilks' theorem [11] that negative twice this difference is asymptotically chi-squared distributed, but the small sample size and the nonstandard parameter $\eta$ bring this approximation into question. The last column in Tables 2 and 3 shows a calibration based on the parametric bootstrap [4]. Using the null-hypothesis parameter estimates, we simulated 249 synthetic data sets (i.e., chromagrams) as if no hot-spots were present. For each data set, we optimized the likelihood under both hypotheses and recorded the difference in maximized loglikelihood. The reported p-values show the proportion of synthetic data sets having more extreme difference than the observed difference. We include the observed data in the numerator and denominator, following Barnard's suggestion [1]. Evidently, the pattern of losses and gains in these bladder cancer tumors is not consistent with sporadic change. Some regions exhibit significantly more change than others.

A critical component in the likelihood calculation is the conditional probability of data at a chromosome given the configuration of hot spots $G$. By Bayes rule, we obtain an expression for the probability of $G$ given the data. This can be marginalized to produce a conditional probability for each chromosome arm: the conditional probability that it is hot given the data, i.e., $P(G_p = +|\text{data})$ and $P(G_q = +|\text{data})$. Maximum-likelihood parameter estimates are plugged in to obtain the numerical values. These are summarized in Figure 2. Figure 2 is modeled after the chromagram in Figure 1, with a vertical axis separating the $p-$arm and the $q-$arm results. Instead of observed rates of change, the shaded area indicates the posterior probability of a hot spot on that arm given the observed counts. This can provide an effective smoothing of the data, extracting significantly elevated rates from background. At the 95% level, chromosome 8p exhibits a significantly elevated loss rate, and chromosomes 5p, 7p, and 20q exhibit significantly elevated gain rates.

## 3. Instability-Selection Modeling.

**3.1. Overview.** In the remainder of this paper we approach the problem of analyzing genome variation from quite a different perspective, although interesting connections to Section 2 do arise. The instability-selection framework introduced in [9] is a two-step plan for building a probability model for data such as those in Figure 1. First we characterize effects of genetic instability in a progenitor cell, and then we apply selection pressures to ask whether or not the clone of descendants of this progenitor will populate an observed tumor. The dual advantages of this approach are that it simplifies the specification process by isolating individual components, and it ensures that the final model respects the basic biological fact of instability and selection. The remaining discussion is devoted to our main result that the linked hot-cold model from Section 2 can be derived within the intability-selection framework. We study deletions to illustrate the argument.

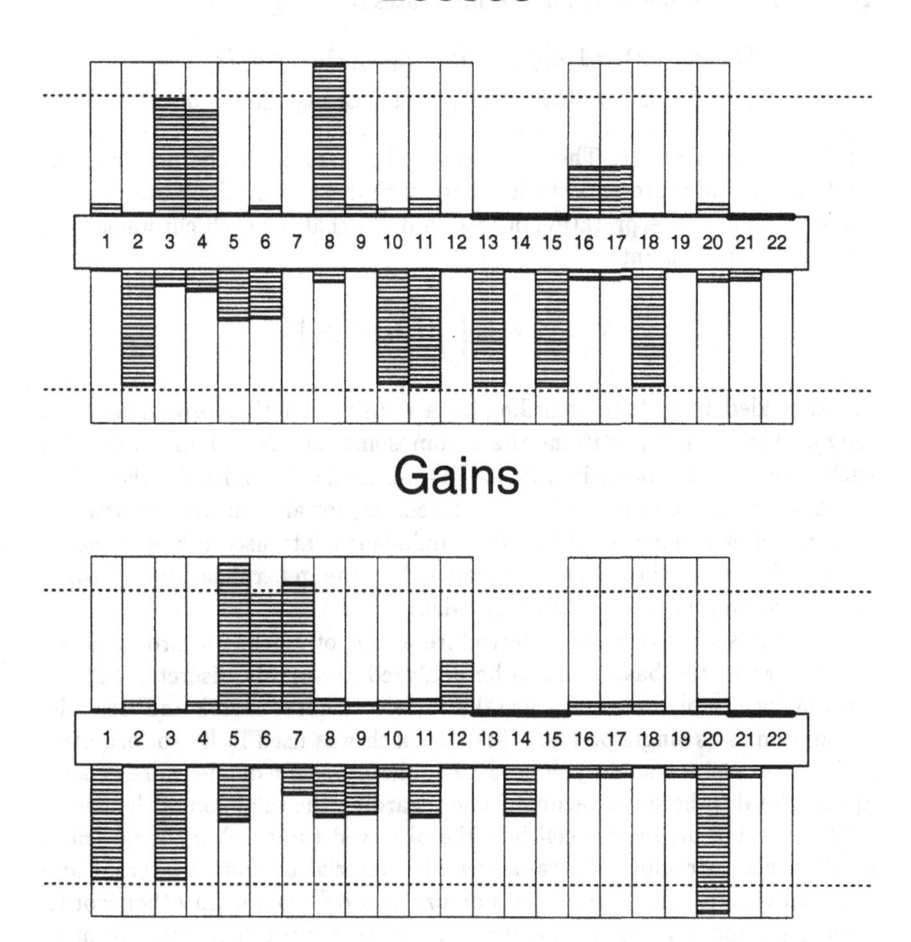


FIG. 2. *Posterior probability of an elevated rate of change:* $P(G_p = +|\text{data})$ *and* $P(G_q = +|\text{data})$. *Reference lines at 95% posterior probability.*

**3.2. Deletions.** Initially, let us focus on a single non-acrocentric chromosome pair in the diploid progenitor cell. A generic position along either chromosome will be denoted $x$, and we consider the chromosome locations to form an interval $[-1,1]$. Along such an idealized chromosome, the centromere occurs at $x = 0$, the $p-$arm is $[-1,0)$ and the $q-$arm is $(0,1]$. We ignore the fact that $p-$arms are physically shorter than $q-$arms in real chromosomes. Genetic instability may create deletions in either chromosome copy, and we encode such deletions with two independent and identically distributed Bernoulli processes $\{L_f(x), L_m(x) : x \in [-1,1]\}$, where the subscript indicates inheritance from the father or the mother. We further consider the common distribution of these processes to be Markovian and governed by the following infinitesimal rates of change:

$$P\{L_i(x+dx)=1|L_i(x)=0\}=\alpha_0\lambda_0(x)dx+o(dx),$$
$$P\{L_i(x+dx)=0|L_i(x)=1\}=(1-\alpha_0)\lambda_0(x)dx+o(dx),$$

as $dx \to 0$ for $i = m, f$. The function $\lambda_0(x)$ is assumed to be positive, and continuous, and represents the intensity of changes in $L_i$ along the chromosome. A simple interpretation of this model is that for each chromosome a number of break points

$$N \sim \text{Poisson}\left(\int_{-1}^{1} \lambda_0(x)\, dx\right)$$

are sprinkled in $[-1, 1]$ according to a density function proportional to $\lambda_0(x)$. This action partitions the chromosome into $N + 1$ intervals. For each interval in the partition, we toss a $\alpha_0$-biased coin to decide whether or not $L(x) = 1$ (deletion) or $L(x) = 0$ (retention) for all $x$ in that interval [3]. The model developed in [9] for allelic-imbalance data used a homogeneous Bernoulli process (i.e., $\lambda_0(x)$ constant), but the generalization is helpful here to develop models for the CGH data.

It may seem excessive to introduce a pair of stochastic processes $L_f$ and $L_m$ when the basic data to be analyzed are highly discrete. Indeed, one may profitably invoke the instability-selection framework and treat the chromosomes as single points. This approach was used in [9] to formulate a simple model for allelotype data. The main benefit of the more general approach taken here is a technical one regarding the selection of the genetically unstable progenitor cell into the observed tumor. Very rarely does deletion act to remove the same genetic material on both maternal and paternal chromosomes (so-called homozygous deletions). In other words, even if $L_f$ and $L_m$ are independent in the progenitor cell, selective pressures will ensure that the cell is not viable if $L_f(x) + L_m(x) = 2$ for some range of $x$, and thus this cell will not have a chance to populate a tumor at the time of observation. The notion is that removal of both copies of a set of critical genes is a disastrous strategy for the progenitor cell. Considering this argument, we say the progenitor cell is viable, denoted VI, if $L(x) = L_f(x) + L_m(x) \leq 1$ for all $x \in [-1, 1]$. The technical benefit is that $L(x)$ in a viable cell retains the Markov Bernoulli structure of the parent processes. Indeed, one may calculate that

$$P\{L(x+dx)=1|L(x)=0,\text{VI}\}=\alpha\lambda(x)dx+o(dx),$$
$$P\{L(x+dx)=0|L(x)=1,\text{VI}\}=(1-\alpha)\lambda(x)dx+o(dx).$$

where $\alpha = 2\alpha_0/(1+\alpha_0)$ and $\lambda(x) = (1+\alpha_0)\lambda_0(x)$. Deletions become encoded by a single process. We consider a simple and useful form for $\lambda(x)$ in Section 4.1.

**3.3. Selection.** The instability-selection framework naturally encodes the idea that the inactivation of certain genes can enhance the probability of selection into a tumor. In essence, this is the definition of a tumor suppressor gene. A tumor suppressor gene is a gene whose products are involved in normal cell cycle regulation, and whose inactivation confers a selective growth advantage. The primary theoretical result in [9] concerns the probability of deletion at a suppressor gene locus in the cells of an observed tumor. Under certain conditions, this probability exceeds the probability of deletion elsewhere in the genome, justifying the identification of regions that exhibit elevated loss rates. Such regions may harbor suppressor genes. Mathematically, parameters governing selection characterize the effects of these putative suppressor genes.

The deletion of a chromosomal region containing a suppressor gene certainly inactivates one copy. We noted earlier that typically we do not observe deletions on both chromosome copies, so the complete inactivation of a suppressor gene requires some other mechanism beyond deletion. One copy may contain an inherited defect, as in Knudson's two hit model [6, 7]. Alternatively, some genetic or epigenetic factor may silence the effect of this gene copy. For example, exposure to a carcinogen could mutate and inactivate that allele. As no data are available on this aspect of the model, we take a pragmatic and simple approach, and assume that non-deleted alleles are inactivated by independent coin tosses, say with inactivation probability $\gamma$. For notation, we say that there has been a loss of suppression, LoS, at a given suppressor gene locus if both alleles are somehow inactivated. Otherwise, there is retention of suppression, RoS.

If genetic instability effects an inactivation of suppressor genes, then the viable cell has an enhanced probability of being selected into a tumor. We denote this even by SEL, i.e. that the descendants of the progenitor cell are selected. Our model must specify precisely how this enhancement occurs. A complicating factor is that suppressor genes may exist on more than one chromosome arm. Perhaps the most simple model of epistasis is

$$\beta_0 = P(\text{SEL}|\text{RoS everywhere}) < P(\text{SEL}|\text{LoS anywhere}) = \beta_1.$$

The three selection parameters $\gamma$, $\beta_0$ and $\beta_1$ are confounded, and we will not be able to estimate them separately from the CGH loss data. In the next section we see how they enter the probability of data.

## 4. A specific model.

**4.1. Instability.** We postulate an intensity function $\lambda(x)$ as in Figure 3, with a very low rate of change on either arm, but with a high rate of change near the centromere $x = 0$. Taking the maximum intensity $h = -\log(\tau)/c$ for some $\tau \in (0, 1)$, the total intensity $\int \lambda(x)\,dx = -\log(\tau)$, no matter how small is $c$. For small $c$, the $N$ breaks, where $N \sim$Poisson$(-\log(\tau))$, all occur near the centromere. In the limit of decreasing $c$, a realization of $\{L(x)\}$ is obtained very simply. We toss a

$\tau$−biased coin to decide whether or not the centromere breaks. We say that *linkage* occurs if the centromere does not break, and thus linkage occurs with probability $\tau$. The outcome determines a partition of the chromosome; either into two separate arms, or the null partition having the whole chromosome by itself. Subsequently, an $\alpha$−biased coin is tossed for each element of the partition to determine whether or not it is deleted. So in the event of linkage, a deletion indicates deletion of the entire chromosome. Deletions of arms are conditionally independent given no linkage.

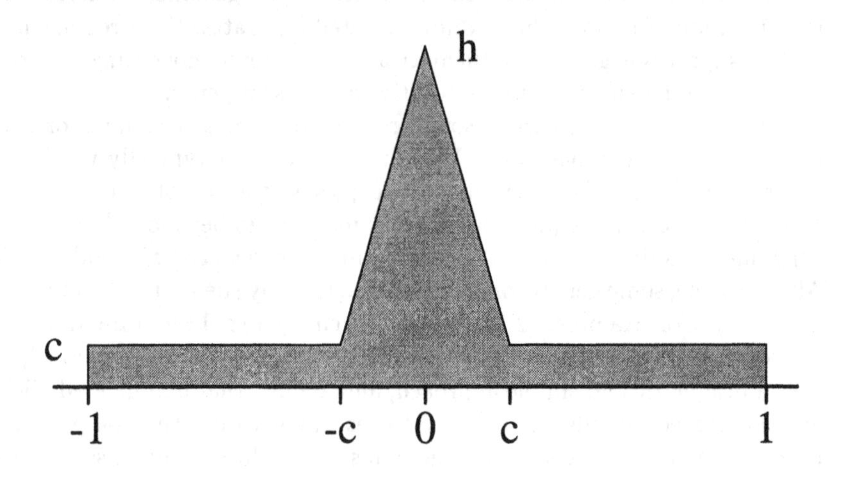


FIG. 3. *Intensity function $\lambda(x)$. Here $h = -\log(\tau)/c$ where $\tau$ is interpreted as the linkage probability.*

With this simple 2-parameter formulation we can calculate the probability of the five basic states, at least for a viable cell prior to selection into a tumor (Table 4). For example, the state $S_1$ of no loss can happen either if there is linkage, and the deletion coin comes up tails, or if there is no linkage and the deletion coin comes up tails twice, giving probability $\tau\bar{\alpha} + \bar{\tau}\bar{\alpha}^2$.

TABLE 4

*Probability of the five observable CGH states in a viable cell prior to selection into a tumor. Note that $\bar{x} = 1 - x$.*

| State $S_i$ | $P(S_i \vert \text{VI})$ |
|---|---|
| $S_1$ =[no loss] | $\bar{\tau}\bar{\alpha}^2 + \tau\bar{\alpha}$ |
| $S_2$ =[tied loss] | $\tau\alpha$ |
| $S_3$ =[$p$-arm loss] | $\bar{\tau}\alpha\bar{\alpha}$ |
| $S_4$ =[$q$-arm loss] | $\bar{\tau}\alpha\bar{\alpha}$ |
| $S_5$ =[untied loss] | $\bar{\tau}\alpha^2$ |

**4.2. Selection.** Model probabilities for observations $S_1, \ldots, S_5$ are calculated as conditional probabilities given selection into the tumor. Noting that viability VI is a prerequisite for selection, SEL $\subset$ VI, we have

$$P(S_i|\text{SEL}) \propto P(S_i|\text{VI})\, P(\text{SEL}|S_i), \tag{4.1}$$

with Table 4 holding probabilities $P(S_i|\text{VI})$. Our assumptions about loss and retention of suppression determine the second factor, tabulated in Table 5. The conditional probability of selection given the phenotype $S_i$ depends on the genotype of normal cells, that is, on whether or not a suppressor gene exists on each chromosome arm. For example, if in normal cells there is a suppressor gene on the $p$−arm and not on the $q$−arm, then

$$\begin{aligned} P(\text{SEL}|S_i) &= P(\text{SEL}|\text{RoS})\, P(\text{RoS}|S_i) + P(\text{SEL}|\text{LoS})\, P(\text{LoS}|S_i) \\ &= \beta_0 P(\text{RoS}|S_i) + \beta_1[1 - P(\text{RoS}|S_i)]. \end{aligned}$$

Retention of suppression, RoS, occurs if at least one copy of the suppressor gene remains intact, and the probability of this depends on the deletion status. For example, if there has been a deletion on the $p$−arm, then LoS occurs with probability $\gamma$, but the chance of LoS is smaller, $\gamma^2$, if no deletion has occurred.

The case of a single suppressor gene locus on the chromosome is particularly simple, and relies on probabilities

$$\begin{aligned} a &= \beta_0 + \gamma^2(\beta_1 - \beta_0) \\ b &= \beta_0 + \gamma(\beta_1 - \beta_0), \end{aligned}$$

satisfying $a < b$. The case of two suppressor gene loci is more complicated, since there are more silencing events to consider. Selection probabilities in these cases are (see Table 5)

$$\begin{aligned} u &= \beta_0 + (\beta_1 - \beta_0)[1 - (1-\gamma)^2(1+\gamma)^2] \\ v &= \beta_0 + (\beta_1 - \beta_0)[1 - (1-\gamma)^2(1+\gamma)] \\ w &= \beta_0 + (\beta_1 - \beta_0)[1 - (1-\gamma)^2], \end{aligned}$$

and satisfy $u < v < w$.

We are now in a position to calculate the probability of CGH states in the observed tumor (4.1). By Bayes rule, these probabilities are formed simply by normalizing the product of entries in Table 4 times those in Table 5. Introducing $\beta = b\alpha/[b\alpha + a\bar{\alpha}] > \alpha$, we uncover a fascinating result, at least for the first three genotype classes. *Observation probabilities $P(S_i|\text{SEL})$ in this instability-selection model equal probabilities in the linked hot-cold model of Section 2.* Thus, Table 1 records the observation probabilities. Hot spots are locations of suppressor genes. The take-home

TABLE 5

*Probability of selection into a tumor from each of the five observable CGH states. Selection probability depends on the configuration of suppressor genes.* $(+/-, +/-)$ *indicate presence or absence on p−arm and q−arm. We use* $a = \beta_0 + \gamma^2(\beta_1 - \beta_0)$, $b = \beta_0 + \gamma(\beta_1 - \beta_0)$, *and* $u$, $v$, *and* $w$ *as defined in the text.*

| | $P(\text{SEL}\|S_i)$ for given normal genotype | | | |
|---|---|---|---|---|
| State $S_i$ | $(-,-)$ | $(+,-)$ | $(-,+)$ | $(+,+)$ |
| $S_1$ | $\beta_0$ | $a$ | $a$ | $u$ |
| $S_2$ | $\beta_0$ | $b$ | $b$ | $w$ |
| $S_3$ | $\beta_0$ | $b$ | $a$ | $v$ |
| $S_4$ | $\beta_0$ | $a$ | $b$ | $v$ |
| $S_5$ | $\beta_0$ | $b$ | $b$ | $w$ |

message is that the simple linked hot-cold model can be derived using primitive assumptions concerning the nature of tumorigenesis within the instability-selection framework.

The case of two suppressor gene loci on a chromosome does not fit exactly, since the $\gamma$ parameter enters $P(\text{SEL}|S_i)$ in a complicated way. The last column of Table 1 is a reasonable approximation, it turns out, especially when $\beta$ is not much larger than $\alpha$. In any event, the case of multiple suppressor genes on a chromosome appears to be unlikely in the CGH data analyzed here.

Among all the parameters used in the derivation, only $\alpha$, $\beta$, and $\tau$ enter the observation probabilities. Parameters $\gamma$, $\beta_0$, and $\beta_1$ governing selection are reduced to the single elevated rate of change $\beta$.

We end with a cautionary note. Changes on different chromosomes are independent according to the linked hot-cold model. This may be a reasonable assumption if selection acts on the chromosome pair. But selection acts on the whole cell, a fact which introduces potential dependencies among changes on different chromosomes in the observed tumor. Details of such dependence remain to be worked out, but initial calculations show that no dependence is induced between chromosomes under the null hypothesis of sporadic loss.

## REFERENCES

[1] G.A. BARNARD, *(In discussion following a paper by Bartlett.)* J. Roy. Statist. Soc. B, **25** (1963), pp.294.

[2] A.D. CAROTHERS, *A likelihood-based approach to the estimation of relative DNA copy number by comparative genomic hybridization.* Biometrics, **53**, (1997), pp. 848-856.

[3] D.J. DALEY AND D. VERE-JONES, *An Introduction to the Theory of Point Processes.* Springer-Verlag, 1988.

[4] B. EFRON AND R. TIBSHIRANI, *An Introduction to the Bootstrap.* Chapman and Hall, 1993.

[5] A. KALLIONIEMI, O-P. KALLIONIEMI, G. CITRO, G. SAUTER, S. DEVRIES, R. KERSCHMANN, P. CAROLL, AND F. WALDMAN *Identification of gains and losses of DNA sequences in primary bladder cancer by comparative genomic hybridization.* Genes, Chromosomes, and Cancer, **12** (1995), pp. 213-219.

[6] A.G. KNUDSON *Mutation and cancer: Statistical study of retinoblastoma.* Proc. Natl. Acad. Sci., **68** (1971), pp. 820-823

[7] S.H. MOOLGAVKAR AND A.G. KNUDSON *Mutation and Cancer: A model for human carcinogenesis.* J. Natl. Cancer Institute, **67** (1981), pp. 16-23.

[8] M.A. NEWTON, S.Q. WU, AND C.A. REZNIKOFF *Assessing the significance of chromosome-loss data: Where are the suppressor genes for bladder cancer?* Statistics in Medicine, **13** (1994), pp. 839-858.

[9] M.A. NEWTON, M.N. GOULD, C.A. REZNIKOFF, AND J.D. HAAG, *On the statistical analysis of allelic-loss data,* Statistics in Medicine, **17** (1998), pp. 1425–1445.

[10] STATISTICAL SCIENCES, *S-PLUS Guide to Statistical and Mathematical Analysis, Version 3.2,* Seattle: StatSci, a division of MathSoft, Inc., 1993.

[11] S.S. WILKS, *The large sample distribution of the likelihood ratio for testing composite hypotheses.* Annals of Mathematical Statistics **9**, (1938), pp. 60-62.

[12] T.R. YEAGER, S. DEVRIES, D.F. JARRARD, C. KAO, S.Y. NAKADA, T.D. MOON, R. BRUSKEWITZ, W.M. STADLER, L.F. MEISNER, K.W. GILCHRIST, M.A. NEWTON, F.M. WALDMAN, AND C.A. REZNIKOFF, *Overcoming cellular senescence in human cancer pathogenesis.* Genes and Development **12**, (1998), pp. 163-174.

## IMA SUMMER PROGRAMS

1987 Robotics
1988 Signal Processing
1989 Robust Statistics and Diagnostics
1990 Radar and Sonar (June 18 - June 29)
New Directions in Time Series Analysis (July 2 - July 27)
1991 Semiconductors
1992 Environmental Studies: Mathematical, Computational, and Statistical Analysis
1993 Modeling, Mesh Generation, and Adaptive Numerical Methods for Partial Differential Equations
1994 Molecular Biology
1995 Large Scale Optimizations with Applications to Inverse Problems, Optimal Control and Design, and Molecular and Structural Optimization
1996 Emerging Applications of Number Theory (July 15 – July 26)
Theory of Random Sets (August 22 – August 24)
1997 Statistics in the Health Sciences
1998 Coding and Cryptography (July 6 – July 18)
Mathematical Modeling in Industry (July 22 – July 31)
1999 Codes, Systems and Graphical Models

## SPRINGER LECTURE NOTES FROM THE IMA:

*The Mathematics and Physics of Disordered Media*
Editors: Barry Hughes and Barry Ninham
(Lecture Notes in Math., Volume 1035, 1983)

*Orienting Polymers*
Editor: J.L. Ericksen
(Lecture Notes in Math., Volume 1063, 1984)

*New Perspectives in Thermodynamics*
Editor: James Serrin
(Springer-Verlag, 1986)

*Models of Economic Dynamics*
Editor: Hugo Sonnenschein
(Lecture Notes in Econ., Volume 264, 1986)

# LIST OF PARTICIPANTS

## Workshop: Week 1: Statistics in the Health Sciences: Genetics July 7 – 11, 1997

| | |
|---|---|
| Chang, Won-Jae | Seoul National University (Mathematics) |
| Chernoff, Herman | Harvard University (Statistics) |
| Efron, Brad | Stanford University (Statistics) |
| Etzel, Carol | Southern Methodist University (Statistical Sciences) |
| Ewens, Warren | University of Pennsylvania (Biology) |
| Feingold, Eleanor | Emory University (Biostatistics) |
| Funo, E. | University of Minnesota (Applied Statistics) |
| Geisser, Seymour | University of Minnesota (Statistics) |
| Geyer, Charles | University of Minnesota (Statistics) |
| Gu, Chi | Washington University (Biostatistics) |
| Guerra, Rudy | Southern Methodist University (Statistical Sciences) |
| Halloran, M. Elizabeth | Emory University (Biostatistics) |
| Holmes, Susan | Cornell University (Biometry) |
| Kaplan, Daniel | Macalester College (Mathematics) |
| Li, Wen-Hsiung | University of Texas, Houston (Center for Demographic and Population Genetics) |
| Mueller, Laurence | University of California-Irvine (Ecology and Evolutionary Biology) |
| Newton, Michael | University of Wisconsin-Madison (Clinical Science Center-K6/446) |
| Painter, Ian | North Carolina State University (Statistics) |
| Seillier-Moiseiwitsch, Francoise | Univ. of North Carolina (Biostatistics) |
| Siegmund, David | Stanford University (Statistics) |
| Simonsen, Katy | North Carolina State Univ. (Statistics) |
| Speed, Terence | University of California-Berkeley (Statistics) |
| Sun, Fengzhu | Emory University (Genetics) |
| Wang, Ming-Dauh | University of Minnesota (Statistics) |
| Waterman, Michael | University of Southern California (Mathematics) |
| Weir, Bruce | North Carolina State University (Statistics) |
| Wijsman, Ellen | University of Washington (Medicine and Biostatistics) |
| Ylvisaker, Don | University of California-Los Angeles (Mathematics) |

**The IMA Volumes in Mathematics and its Applications**

***Current Volumes:***

23 **Signal Processing Part II: Control Theory and Applications of Signal Processing** L. Auslander, F.A. Grünbaum, J.W. Helton, T. Kailath, P. Khargonekar, and S. Mitter (eds.)
24 **Mathematics in Industrial Problems, Part 2** A. Friedman
25 **Solitons in Physics, Mathematics, and Nonlinear Optics** P.J. Olver and D.H. Sattinger (eds.)
26 **Two Phase Flows and Waves** D.D. Joseph and D.G. Schaeffer (eds.)
27 **Nonlinear Evolution Equations that Change Type** B.L. Keyfitz and M. Shearer (eds.)
28 **Computer Aided Proofs in Analysis** K. Meyer and D. Schmidt (eds.)
29 **Multidimensional Hyperbolic Problems and Computations** A. Majda and J. Glimm (eds.)
30 **Microlocal Analysis and Nonlinear Waves** M. Beals, R. Melrose, and J. Rauch (eds.)
31 **Mathematics in Industrial Problems, Part 3** A. Friedman
32 **Radar and Sonar, Part I** R. Blahut, W. Miller, Jr., and C. Wilcox
33 **Directions in Robust Statistics and Diagnostics: Part I** W.A. Stahel and S. Weisberg (eds.)
34 **Directions in Robust Statistics and Diagnostics: Part II** W.A. Stahel and S. Weisberg (eds.)
35 **Dynamical Issues in Combustion Theory** P. Fife, A. Liñán, and F.A. Williams (eds.)
36 **Computing and Graphics in Statistics** A. Buja and P. Tukey (eds.)
37 **Patterns and Dynamics in Reactive Media** H. Swinney, G. Aris, and D. Aronson (eds.)
38 **Mathematics in Industrial Problems, Part 4** A. Friedman
39 **Radar and Sonar, Part II** F.A. Grünbaum, M. Bernfeld, and R.E. Blahut (eds.)
40 **Nonlinear Phenomena in Atmospheric and Oceanic Sciences** G.F. Carnevale and R.T. Pierrehumbert (eds.)
41 **Chaotic Processes in the Geological Sciences** D.A. Yuen (ed.)
42 **Partial Differential Equations with Minimal Smoothness and Applications** B. Dahlberg, E. Fabes, R. Fefferman, D. Jerison, C. Kenig, and J. Pipher (eds.)
43 **On the Evolution of Phase Boundaries** M.E. Gurtin and G.B. McFadden
44 **Twist Mappings and Their Applications** R. McGehee and K.R. Meyer (eds.)
45 **New Directions in Time Series Analysis, Part I** D. Brillinger, P. Caines, J. Geweke, E. Parzen, M. Rosenblatt, and M.S. Taqqu (eds.)

46 **New Directions in Time Series Analysis, Part II**
D. Brillinger, P. Caines, J. Geweke, E. Parzen, M. Rosenblatt, and M.S. Taqqu (eds.)

47 **Degenerate Diffusions**
W.-M. Ni, L.A. Peletier, and J.-L. Vazquez (eds.)

48 **Linear Algebra, Markov Chains, and Queueing Models**
C.D. Meyer and R.J. Plemmons (eds.)

49 **Mathematics in Industrial Problems, Part 5** A. Friedman

50 **Combinatorial and Graph-Theoretic Problems in Linear Algebra**
R.A. Brualdi, S. Friedland, and V. Klee (eds.)

51 **Statistical Thermodynamics and Differential Geometry of Microstructured Materials**
H.T. Davis and J.C.C. Nitsche (eds.)

52 **Shock Induced Transitions and Phase Structures in General Media** J.E. Dunn, R. Fosdick, and M. Slemrod (eds.)

53 **Variational and Free Boundary Problems**
A. Friedman and J. Spruck (eds.)

54 **Microstructure and Phase Transitions**
D. Kinderlehrer, R. James, M. Luskin, and J.L. Ericksen (eds.)

55 **Turbulence in Fluid Flows: A Dynamical Systems Approach**
G.R. Sell, C. Foias, and R. Temam (eds.)

56 **Graph Theory and Sparse Matrix Computation**
A. George, J.R. Gilbert, and J.W.H. Liu (eds.)

57 **Mathematics in Industrial Problems, Part 6** A. Friedman

58 **Semiconductors, Part I**
W.M. Coughran, Jr., J. Cole, P. Lloyd, and J. White (eds.)

59 **Semiconductors, Part II**
W.M. Coughran, Jr., J. Cole, P. Lloyd, and J. White (eds.)

60 **Recent Advances in Iterative Methods**
G. Golub, A. Greenbaum, and M. Luskin (eds.)

61 **Free Boundaries in Viscous Flows**
R.A. Brown and S.H. Davis (eds.)

62 **Linear Algebra for Control Theory**
P. Van Dooren and B. Wyman (eds.)

63 **Hamiltonian Dynamical Systems: History, Theory, and Applications**
H.S. Dumas, K.R. Meyer, and D.S. Schmidt (eds.)

64 **Systems and Control Theory for Power Systems**
J.H. Chow, P.V. Kokotovic, R.J. Thomas (eds.)

65 **Mathematical Finance**
M.H.A. Davis, D. Duffie, W.H. Fleming, and S.E. Shreve (eds.)

66 **Robust Control Theory** B.A. Francis and P.P. Khargonekar (eds.)

67 **Mathematics in Industrial Problems, Part 7** A. Friedman

68 **Flow Control** M.D. Gunzburger (ed.)

69 **Linear Algebra for Signal Processing**
A. Bojanczyk and G. Cybenko (eds.)
70 **Control and Optimal Design of Distributed Parameter Systems**
J.E. Lagnese, D.L. Russell, and L.W. White (eds.)
71 **Stochastic Networks** F.P. Kelly and R.J. Williams (eds.)
72 **Discrete Probability and Algorithms**
D. Aldous, P. Diaconis, J. Spencer, and J.M. Steele (eds.)
73 **Discrete Event Systems, Manufacturing Systems, and Communication Networks**
P.R. Kumar and P.P. Varaiya (eds.)
74 **Adaptive Control, Filtering, and Signal Processing**
K.J. Åström, G.C. Goodwin, and P.R. Kumar (eds.)
75 **Modeling, Mesh Generation, and Adaptive Numerical Methods for Partial Differential Equations** I. Babuska, J.E. Flaherty, W.D. Henshaw, J.E. Hopcroft, J.E. Oliger, and T. Tezduyar (eds.)
76 **Random Discrete Structures** D. Aldous and R. Pemantle (eds.)
77 **Nonlinear Stochastic PDEs: Hydrodynamic Limit and Burgers' Turbulence** T. Funaki and W.A. Woyczynski (eds.)
78 **Nonsmooth Analysis and Geometric Methods in Deterministic Optimal Control** B.S. Mordukhovich and H.J. Sussmann (eds.)
79 **Environmental Studies: Mathematical, Computational, and Statistical Analysis** M.F. Wheeler (ed.)
80 **Image Models (and their Speech Model Cousins)**
S.E. Levinson and L. Shepp (eds.)
81 **Genetic Mapping and DNA Sequencing**
T. Speed and M.S. Waterman (eds.)
82 **Mathematical Approaches to Biomolecular Structure and Dynamics**
J.P. Mesirov, K. Schulten, and D. Sumners (eds.)
83 **Mathematics in Industrial Problems, Part 8** A. Friedman
84 **Classical and Modern Branching Processes**
K.B. Athreya and P. Jagers (eds.)
85 **Stochastic Models in Geosystems**
S.A. Molchanov and W.A. Woyczynski (eds.)
86 **Computational Wave Propagation**
B. Engquist and G.A. Kriegsmann (eds.)
87 **Progress in Population Genetics and Human Evolution**
P. Donnelly and S. Tavaré (eds.)
88 **Mathematics in Industrial Problems, Part 9** A. Friedman
89 **Multiparticle Quantum Scattering With Applications to Nuclear, Atomic and Molecular Physics** D.G. Truhlar and B. Simon (eds.)
90 **Inverse Problems in Wave Propagation** G. Chavent, G. Papanicolau, P. Sacks, and W.W. Symes (eds.)
91 **Singularities and Oscillations** J. Rauch and M. Taylor (eds.)

92 **Large-Scale Optimization with Applications, Part I: Optimization in Inverse Problems and Design**
L.T. Biegler, T.F. Coleman, A.R. Conn, and F. Santosa (eds.)
93 **Large-Scale Optimization with Applications, Part II: Optimal Design and Control**
L.T. Biegler, T.F. Coleman, A.R. Conn, and F. Santosa (eds.)
94 **Large-Scale Optimization with Applications, Part III: Molecular Structure and Optimization**
L.T. Biegler, T.F. Coleman, A.R. Conn, and F. Santosa (eds.)
95 **Quasiclassical Methods**
J. Rauch and B. Simon (eds.)
96 **Wave Propagation in Complex Media**
G. Papanicolaou (ed.)
97 **Random Sets: Theory and Applications**
J. Goutsias, R.P.S. Mahler, and H.T. Nguyen (eds.)
98 **Particulate Flows: Processing and Rheology**
D.A. Drew, D.D. Joseph, and S.L. Passman (eds.)
99 **Mathematics of Multiscale Materials** K.M. Golden, G.R. Grimmett, R.D. James, G.W. Milton, and P.N. Sen (eds.)
100 **Mathematics in Industrial Problems, Part 10** A. Friedman
101 **Nonlinear Optical Materials** J.V. Moloney (ed.)
102 **Numerical Methods for Polymeric Systems** S.G. Whittington (ed.)
103 **Topology and Geometry in Polymer Science** S.G. Whittington, D. Sumners, and T. Lodge (eds.)
104 **Essays on Mathematical Robotics** J. Baillieul, S.S. Sastry, and H.J. Sussmann (eds.)
105 **Algorithms For Parallel Processing** M.T. Heath, A. Ranade, and R.S. Schreiber (eds.)
106 **Parallel Processing of Discrete Problems** P.M. Pardalos (ed.)
107 **The Mathematics of Information Coding, Extraction, and Distribution** G. Cybenko, D.P. O'Leary, and J. Rissanen (eds.)
108 **Rational Drug Design** D.G. Truhlar, W. Howe, A.J. Hopfinger, J. Blaney, and R.A. Dammkoehler (eds.)
109 **Emerging Applications of Number Theory** D.A. Hejhal, J. Friedman, M.C. Gutzwiller, and A.M. Odlyzko (eds.)
110 **Computational Radiology and Imaging: Therapy and Diagnostics**
C. Börgers and F. Natterer (eds.)
111 **Evolutionary Algorithms** L.D. Davis, K.De Jong, M.D. Vose, and L.D. Whitley (eds.)
112 **Statistics in Genetics** M.E. Halloran and S. Geisser (eds.)
113 **Grid Generation and Adaptive Algorithms** M.W. Bern, J.E. Flaherty, M. Luskin (eds.)
114 **Diagnosis and Prediction** S. Geisser (ed.)
115 **Pattern Formation in Continuous and Coupled Systems: A Survery Volume**
M. Golubitsky, D. Luss, S.H. Strogatz (eds.)